Confined Space Rescue

George J. Browne

Gus S. Crist

Confined
Space
Rescue

George J. Browne

Gus S. Crist

Delmar Publishers

I(T)P® **International Thomson Publishing**

Albany • Bonn • Boston • Cincinnati • Detroit • London • Madrid
Melbourne • Mexico City • New York • Pacific Grove • Paris • San Francisco
Singapore • Tokyo • Toronto • Washington

NOTICE TO THE READER

Cover Design: Brucie Rosch

Delmar Staff:
Publisher: Alar Elken
Acquisitions Editor: Mark Huth
Developmental Editor: Jeanne Mesick
Project Editor: Barbara Diaz
Production Manager: Mary Ellen Black
Editorial Assistant: Dawn Daugherty

COPYRIGHT © 1999
By Delmar Publishers
an International Thomson Publishing company

The ITP logo is a trademark under license
Printed in the United States of America

For more information contact:

Delmar Publishers
3 Columbia Circle, Box 15015
Albany, New York 12212-5015

International Thomson Publishing Europe
Berkshire House
168-173 High Holborn
London, WC1V7AA
United Kingdom

Nelson ITP, Australia
102 Dodds Street
South Melbourne,
Victoria, 3205 Australia

Nelson Canada
1120 Birchmont Road
Scarborough, Ontario
M1K 5G4, Canada

International Thomson Publishing France
Tour Maine-Montparnasse
33 Avenue du Maine
75755 Paris Cedex 15, France

International Thomson Editores
Seneca 53
Colonia Polanco
11560 Mexico D. F. Mexico

International Thomson Publishing GmbH
Königswinterer Straße 418
53227 Bonn
Germany

International Thomson Publishing Asia
60 Albert Street
#15-01 Albert Complex
Singapore 189969

International Thomson Publishing Japan
Hirakawa-cho Kyowa Building, 3F
2-2-1 Hirakawa-cho, Chiyoda-ku,
Tokyo 102, Japan

ITE Spain/Paraninfo
Calle Magallanes, 25
28015-Madrid, Espana

2 3 4 5 6 7 8 9 10 XXX 03 02

Library of Congress Cataloging-in-Publication Data
Browne, George, 1949-
    Confined space rescue / George Browne and Gus Crist.
        p.   cm.
    ISBN 0-8273-8559-5 (sc)
    1. Industrial safety.   2. Rescue work.   3. Industrial accidents,
I. Crist, Gus.   II. Title.
T55.B754   1998
628.9'2—dc21   98-42594
                                                              CIP

This book is dedicated not only to all those people who believe that assisting their fellow man can make a difference, but to their families who stand behind them.

And to our families who knew we were trying to do something special, thanks.

Kathy, Helene, Samuel, Amy, and Zach

Cheryl, Tim, Phil and Kieran

PREFACE

How many times have you seen headlines or heard a reporter describe a dramatic rescue with the words *brave* and *courageous*? This description brings to mind sensational scenes of danger, human drama, and the successful rescue of an otherwise perishing victim. Our valiant rescuers are viewed as having immeasurable courage and stand before the public trying to put the rescue into modest, human terms. There is no doubt that very often the emergency services are called upon to perform rescues which place the rescuers at risk to protect and save a victim. There is no doubt that it takes a certain level of courage, bravery, or guts to choose to place yourself at risk to attempt that rescue. However, if that was all it took, we would truly cheapen the efforts of all emergency workers. The bravado, the apparent fearlessness, and the gallantry that the public sees is only a small part when compared with the level of commitment and competency that must be developed to even attempt a rescue. Emergency workers do not foolishly place themselves at risk. Instead, they prepare, train, plan, and train some more for the situations they expect to face. That fact is what this book is about——preparing yourself for a new challenge.

As this book was being written, the National Fire Protection Association was developing a standard (NFPA 1670) to define the levels of competency for people involved in different types of technical rescue, not only confined-space rescue. We wish the standard had been completed and adopted before the book was completed, but unfortunately this was not the case. That does not mean the proposed NFPA standard and this book are at odds; in fact, we wrote this book while constantly glancing at the proposed standard and the different drafts of it. It means that, as the standard evolves, so must your confined-space rescue organization and operations. When the NFPA standard comes out, you should view it as a tool to help you better define the level of response and the level of training that you want to achieve in regard to understanding technical rescue. Defining your level of response capability and then working to attain that level will be part of what makes your rescue team successful. The backbone to your rescue efforts will be commitment, competency, and confidence. Commitment is something that each of us must develop internally and personally, and when enough people with the same level of commitment get together, you develop a team that knows where it wants to go and what it wants to do when it gets there. Competency comes from training, testing your skills and equipment, and knowing your limits in regard to skills and equipment. Confidence comes from having been in the same or similar situation before, either through training or practical experience, and knowing that you and your equipment can perform according to plan. Confidence also requires trust. Trust not only in yourself and in your equipment, but also in the people with whom you work. How many people would have the confidence to enter a confined space for rescue if they did not believe that the other members of the rescue team could get them out of that space?

We have briefly mentioned NFPA standards, and it should be pointed out that more than one NFPA standard will influence how you train, what equipment you use, and how you maintain it. In addition to the NFPA are other standard-making organizations, some voluntary and some regulatory. Whether you elect to follow a voluntary standard or are required to follow a regulatory standard, you are gaining reliability in your methods of operating and in performance of your equipment. Standards that prescribe how a particular piece of equipment is to be designed are based on how the equipment is expected to be used. Based on that design, you can expect a high degree of reliability when you use it as intended; but that degree of reliability can be lost without a program of maintenance and inspection. Reliability does not only include equipment; the personnel that are the heart of your rescue operation must also be reliable. Just as equipment requires maintenance and inspection, so do rescue personnel. For people, maintenance and inspection must include periodic training and testing of their skills in practical exercises. At times, your maintenance and inspection will reveal problems that you must address. If you cannot repair the problem, then you must work within the parameters that have been delivered to you.

Sometimes, however, because of a lack of information, people do not recognize that there is a problem. Resources might be used improperly, poorly maintained, or they might not be prepared for what is expected of them. The question then becomes, Is that how you want to operate? The old cliché, "A little knowledge is a dangerous thing," really rings true in confined-space rescue. People who are in charge of a rescue and are poorly prepared, either through knowledge or training, must be held accountable if someone is needlessly placed at risk. Similarly, subordinates who blindly follow an unqualified leader must share in that responsibility, especially if the subordinates had the training and the knowledge to know better. This statement is not designed to encourage freelancing or mutiny at the emergency scene, but rather states that you have an obligation to prepare, in advance, for this type of emergency if you are going to become involved in confined-space rescue. If you fail to plan, you plan to fail!

As emergency responders, most of us got involved because we felt that we could help people, that we could make a difference. We cannot successfully help people unless we have prepared in advance. Take the time to define the confined-space rescue problem occurring in your response community, determine what you want to do about it in terms of preparing for response and/or developing outside resources, train for that response, prepare standard operating procedures as guidelines to assist you, and then evaluate how your plan is working. No emergency organization can ever prepare to handle every type of emergency on its own, but it can plan for those times when it needs assistance. Most importantly, know your limits! There is nothing to be gained by injuring or killing rescuers in a futile attempt to rescue a victim that you cannot reach. Professionals do not take chances, they minimize the risks!

What we do as rescuers has value to society and is an admirable profession. We do make a difference and people's lives are better for it, but that is its own reward. Stay safe.

ACKNOWLEDGMENTS

The authors and Delmar Publishers wish to acknowledge and thank the members of our editorial panel for their help in developing the concept of this series and reviewing the manuscript.

Steve Fleming
National Association for Search and Rescue
Fort Collins, CO

Christopher J. Naum
L.A. Emergency Management & Training Consultant
Liverpool, NY

Mike Brown
Virginia Beach Fire Department
Virginia Beach, VA

Robert Hancock
Hillsborough County Fire Rescue
Tampa, FL

Jerry Nulliner
Fishers Fire Department
Fishers, IN

Rod Westerfield
Training Horizons
Hollister, MO

Jeffrey Reames
Western Wisconsin Technical College
LaCrosse, WI

Ben Blankenship
El Dorado Fire Department
El Dorado, AR

Mike Flavin
St. Louis County Fire Academy
St. Louis, MO

Bill Shouldis
Philadelphia Fire Department
Philadelphia, PA

John J. Griffin
Holden, MA

Don Krain, Ken Toth, and Chris Galamb
GATX Terminals Co.
Carteret, NJ

Marc Brodt
Colgate Palmolive Co.
New York, NY

John Gluchowski & Ken Tardell
Stepan Co.
Maywood, NJ

The members of the Weehawken Fire Department
Weehawken, NJ

Gregory A. DePaul & James S. Curley
Berkeley Township Emergency Response Team
Bayville, NJ

Robert Hansson
Hazardous Materials Emergency Planning Unit
New Jersey State Police
West Trenton, NJ

Pinewald Pioneer Technical Rescue Team
Station #20
Pinewald, NJ

All Industrial Safety Products, Inc.
Edgewater Park, NJ

Olympic Glove
Elmwood Park, NJ

Uniquest, Inc.
Consultants and Training Professionals
Toms River, NJ

Ocean County Utilities Authority
Ocean County, NJ

We also gratefully acknowledge the assistance of Warren Carr, Jr., of
the Latham Fire Department, Latham, NY, for assisting us with the
photography for this book.

C O N T E N T S

CHAPTER

1 Confined Spaces and Their Hazards

OBJECTIVES

After completing this chapter, the reader should be able to define the following terms:

- confined spaces
- atmospheric hazards
- flammable atmospheres
- oxygen-deficient atmospheres
- oxygen-enriched atmospheres
- toxic atmospheres
- physical hazards
- engulfment
- permit-required confined spaces

and define the following basic chemical and physical properties:

- flash point
- flammable (explosive) range
- vapor density

INTRODUCTION

A father and his two sons decide to spend a Saturday repairing one of the waste tanks on the septic system for their house. The father and one son use a ladder to enter the tank and begin working while the other son remains outside to haul out buckets of dry waste. After working for a short time, the son outside the tank notices that his father and brother have collapsed inside the tank. He immediately goes inside the house and calls 911. When the son comes back, he enters the tank opening to assist his father and brother. In the end, the father and both sons die before they can be rescued.

This incident is based on an actual confined-space accident. Unfortunately it is a typical confined-space accident, with multiple deaths, including a rescuer. During the 1980s, the **National Institute for Occupational Safety and Health (NIOSH)** studied eight cases involving deaths that occurred in confined spaces. There were sixteen fatalities and fifty-three injuries; of those sixteen fatalities, ten were would-be rescuers. How many of us, as rescue personnel, might find ourselves in that same position?

DEFINING CONFINED SPACES

What is a confined space? What about the space makes it so potentially hazardous? How can we safely operate in a confined space? To begin, the **Occupational Safety and Health Administration (OSHA)** defines a confined space as that which:

1. is large enough and so configured that an employee can bodily enter and perform assigned work (as shown in Figures 1–1 and 1–2).
2. has limited or restricted means for entry or exit (for example, tanks, vessels, silos, storage bins, hoppers, vaults, and pits are spaces that may have limited means of entry); and
3. is not designed for continuous employee occupancy.

Each of these items is important because each helps to define the problems with confined spaces. The fact that a confined space is large enough to enter and to work in means that a person can enter the space and be exposed to any hazards it may contain.

A worker would have to physically leave or be removed from the space to be out of danger from the hazards contained there.

Complicating both entry and exit from a confined space is its limited opening, which can be as small as 18 inches diameter. People entering the space, including rescuers, must try to fit through the opening. If a worker or rescuer has to wear a **self-contained breathing apparatus (SCBA)** to safely enter and work in the space, he may find it difficult to fit through that small opening. In these cases, workers or rescuers may enter the space without using the SCBA, intending to put on the equipment after they have entered. Those entering the space without their SCBA are exposed to whatever atmospheric hazards exist and may fall victim to these hazards before they can put on their SCBA. This situation often occurs when the

▶ **National Institute for Occupational Safety and Health (NIOSH)**

the agency within the U.S. Department of Health and Human Services that identifies work-related diseases and injuries and the potential hazards of new work-related technologies and practices.

▶ **Occupational Safety and Health Administration (OSHA)**

the federal agency within the U.S. Department of Labor that is responsible for creating and enforcing workplace safety and health regulations.

NOTE: The fact that a confined space is large enough to enter and to work in means that a person can enter the space and be exposed to any hazards it may contain.

▶ **self-contained breathing apparatus (SCBA)**

a form of respiratory protection in which the self-contained air supply and related equipment is attached to the wearer and the pressure within the facepiece is greater than the surrounding atmospheric pressure.

FIGURE 1-1

Interior of a confined space. This space is large enough for a person to enter and work in.

FIGURE 1-2

View of a limited opening on a confined space—opening is on right side of boiler.

NOTE: The sense of urgency many of us feel during rescue work blinds us to the true nature of the hazards and can allow us to become victims.

rescuer's attempt is poorly planned. The sense of urgency many of us feel during rescue work blinds us to the true nature of the hazards and can allow us to become victims.

In considering that a confined space is not meant for continuous human occupancy, we must realize that conditions inside the space can change without anyone being there to notice. Certainly a confined space that contains hazardous materials or chemical processes is also expected to contain the hazards of those materials, but what about a space as simple as an underground storm sewer? Runoff from the road, decomposing organic material, and exhaust gases from cars can introduce hazards without anyone being aware of them.

Ultimately, the problems with confined-space entry for normal work practices, such as maintenance and for rescue operations stem from lack of preparation before entry. The situation can easily contain

NOTE: Preparation prior to entry is a necessity.

► **atmospheric hazards**

conditions which present an atmosphere that can be toxic, flammable, oxygen deficient, oxygen enriched, or that obscures visibility.

► **OSHA, Standard 1910.146 Permit-required confined spaces**

the specific section of the code of Federal Regulations which regulates confined spaces and the manner in which activities can occur within those spaces.

NOTE: Airborne combustible dust at a concentration that meets or exceeds its LFL may be approximated as a condition in which the dust obscures vision at a distance of 5 feet (1.52 m) or less.

NOTE: For air contaminants for which OSHA has not determined a dose or permissible exposure limit, other sources of information, such as Material Safety Data Sheets . . ., published information, and internal documents can provide guidance in establishing acceptable atmospheric conditions.

unknown and complicated problems. Only an awareness of the potential hazards and a plan to work within the limits set by the hazards can lead to a safe operation. Therefore, preparation prior to entry is a necessity.

HAZARD RECOGNITION

Atmospheric Hazards

The atmosphere within a confined space may be one of the most critical elements affecting confined space operations. **Atmospheric hazards** in confined spaces may include toxic gases, flammable gases, and oxygen deficiency. You cannot simply look into the space to determine if an atmospheric hazard exists. You cannot effectively detect the presence of a hazardous atmosphere by sniffing it (that alone can kill you). At one time, coal miners used canaries to detect a hazardous atmosphere: Because the birds are so small, a smaller dose of coal gas would kill the canary than would kill a person. A dead canary provided a warning that coal gas was in the area, and the miners could thus escape. Another monitoring device used in the past was a candle, which, would be lowered into a confined space. If the candle went out, that would indicate an oxygen-deficient atmosphere. Imagine what would have happened instead if that space had contained a flammable atmosphere? Fortunately, today's monitoring devices are much more sophisticated, safe, and accurate than canaries or candles.

OSHA, in Standard 1910.146 Permit-required confined spaces, defines a hazardous atmosphere as an *atmosphere that may expose employees to the risk of death, incapacitation, impairment of ability to self-rescue* (i.e., escape unaided from a permit space), *injury, or acute illness from one or more of the following causes:*

1. flammable gas, vapor, or mist in excess of 10 percent of its lower flammable limit (LFL);

2. airborne combustible dust at a concentration that meets or exceeds its LFL; NOTE: This concentration may be approximated as a condition in which the dust obscures vision at a distance of 5 feet (1.52 m) or less.

3. atmospheric oxygen concentration below 19.5 percent or above 23.5 percent;

4. atmospheric concentration of any substance . . . which could result in employee exposure in excess of its dose or permissible exposure limit;

5. any other atmospheric condition that is immediately dangerous to life or health. NOTE: For air contaminants for which OSHA has not determined a dose or permissible exposure limit, other sources of information, such as Material Safety Data Sheets . . ., published information, and internal documents can provide guidance in establishing acceptable atmospheric conditions.

Flammable atmospheres can exist when gases, liquids, dusts, and vapors are introduced into a confined space. These materials can enter the space as part of normal process operations; they may have been introduced into the space by work being performed there or by accident.

In understanding the hazards of a flammable atmosphere within a confined space, it is important to understand certain basic principles about the fuel.

The amount of fuel necessary to support combustion will vary with the fuel. First, sufficient fuel, oxygen, and heat must be present for combustion to occur. When working with fuel vapors, and certain liquids and solids, we must understand the range which indicates if there will be enough fuel to burn. This range is called the **flammable or explosive range,** as shown in Figure 1–3. The flammable range has a lower and an upper flammable limit (also called lower explosive and upper explosive limits). These limits are the minimum and maximum concentrations of the fuel in air needed to support combustion. If the mixture is below the lower flammable limit (LFL), it is considered too lean to burn. If the mixture is above the upper flammable limit (UFL), it is considered too rich to burn. If the temperature of the fuel and air mixture is increased, the flammable range will also increase, whereas if the temperature is decreased, the flammable range will also decrease.

In discussing flammable limits, we must also realize that certain liquids produce enough vapors at normal atmospheric temperatures and pressures that the presence of these liquids within a confined space can create a flammable mixture. Normally these liquids will have a **flash point,** as shown in Figure 1–4, at or below the ambient conditions of the space. The flash point of a material is the temperature at which the liquid gives off enough vapors to be ignited. The determination of the flash point of a material is made in a laboratory and can vary slightly depending on the method used. When the temperature of a liquid is at or above its flash point, it is at least at its lower flammable limit.

NOTE: Sufficient fuel, oxygen, and heat must be present for combustion to occur.

▶ flammable or explosive range

a definite concentration of flammable vapors in air, for a particular material, at which combustion will occur. There is a lower flammable limit and an upper flammable limit at which the vapors are either too lean to burn or too rich.

▶ flash point

the minimum temperature at which a material will produce enough vapors, in air, to form an ignitable mixture near the surface of the material.

FIGURE 1-3

Illustration showing flammable range. The fuel mixture to the left of the LFL is too lean (not enough fuel) and the fuel mixture to the right of the UFL is too rich (too much fuel, not enough oxygen).

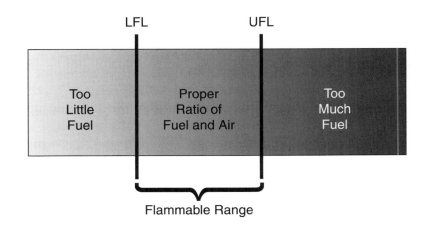

LFL   UFL

| Too Little Fuel | Proper Ratio of Fuel and Air | Too Much Fuel |

Flammable Range

FIGURE 1-4

Different materials have different flash points. Knowing the identity of the material, the flash point, and the temperature of the material will give you an idea of the fire hazard.

Flash Points of Various Liquids

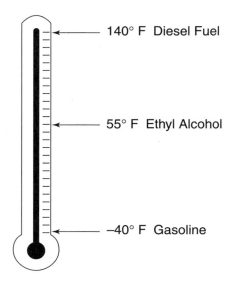

140° F Diesel Fuel

55° F Ethyl Alcohol

−40° F Gasoline

| NOTE: If contained in a confined space, flash fire can create a rapid buildup of pressure which is, in effect, an explosion. It is important to note that certain materials not normally considered combustible can ignite when found as dust.

| NOTE: It is important to realize that, as the flammable range of the materials varies, so does the potential hazard presented by the material.

▶ **calibrated**

the condition of a measuring instrument after its graduations have been checked or corrected.

In dealing with the flammable range of solids, the primary concern is with dusts. Dusts are minute particles of a solid. Because these particles are so finely divided, they are easily heated by an ignition source. Once heated, these dusts ignite and burn. If contained in a confined space, this flash fire can create a rapid buildup of pressure which is, in effect, an explosion. It is important to note that certain materials not normally considered combustible can ignite when found as dust.

To categorize the hazard presented by the presence of a combustible dust, OSHA uses as an *approximation* the condition when the dust obscures vision at a distance of 5 feet (1.52 m) or less. Although this is an approximation, the presence of a dust condition of this magnitude should be cause for concern.

The flammable range for all materials is not identical. Gasoline has a given flammable range of 1.4 to 7.6 percent in air, while anhydrous ammonia has a flammable range of 15.5 to 27 percent in air, as shown in Figure 1–5. Acetylene has a flammable range given as 2.5 to 100 percent in air under certain conditions. It is important to realize that, as the flammable range of the materials varies, so does the potential hazard presented by the material.

By using a limit of 10 percent of the LFL, as given in the OSHA definition, you will be accommodating the variations in materials. Considering a 10 percent LFL action limit provides a greater range of safety in that most metering equipment for measuring combustible gases is **calibrated** on a single gas. This calibration gas has a specific LFL, and staying below 10 percent accommodates other gases which may be present but unknown and have a lower LFL than the gas used for calibrating the meter.

As discussed, a space must have adequate oxygen for combustion to occur and, more importantly, for the support of human life. Normally, the atmospheric air we breathe contains approximately 21 percent oxygen. Oxygen concentrations below 19.5 percent are con-

FIGURE 1-5

Flammable ranges of various materials. Some materials have a very narrow flammable range, while others have a very broad range.

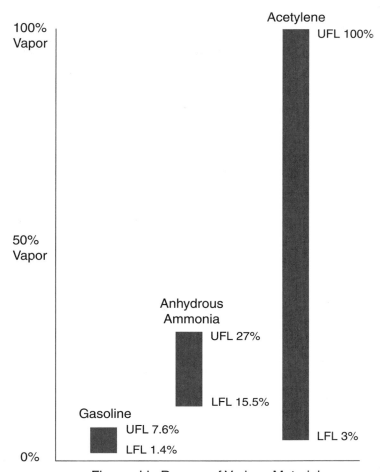

Flammable Ranges of Various Materials

SAFETY
Oxygen concentrations below 19.5 percent are considered oxygen deficient and pose a danger to human life.

NOTE: Oxygen concentrations above 23.5 percent are considered oxygen-enriched.

sidered oxygen deficient and pose a danger to human life, as shown in Figure 1–6. In addition, low oxygen concentrations affect the flammability of fuels within the space. It is entirely possible that a potentially explosive atmosphere can be present, needing only an increase in oxygen to move into the explosive range. Introducing oxygen into the confined space could possibly bring together all ingredients needed for an explosion. Oxygen concentrations above 23.5 percent are considered oxygen-enriched. Oxygen-enriched atmospheres pose a hazard in that they can rapidly accelerate combustion. Additionally, materials that are not easily ignited in normal air or are not normally considered combustible can be ignited and thus burn in an oxygen-rich atmosphere. Oxygen levels within the confined space can be affected by what occurred in it, prior to entry, by what work was being done in it, or even by sources outside of it.

Any discussion of atmospheric hazards of confined spaces would not be complete without considering the presence of hazardous materials. Hazardous materials can be present by many different means. The simplest source of hazardous materials originates from those stored or processed in the confined space. Other sources include

FIGURE 1-6

Effects of varying levels of oxygen on people.

Effects of Reduced O_2

- ❖ 21%—normal
- ❖ 19.5%—OSHA definition as oxygen-deficient
- ❖ 17%—some muscular impairment, increased respiratory rate
- ❖ 12%—dizziness, headache, rapid fatigue
- ❖ 9%—unconsciousness
- ❖ 6%—death within a few minutes

---

items or processes brought into the area by the workers performing work there. Still other sources might be chemical reactions occurring in the space such as the decomposition of organic material either in the area or in the ground surrounding it.

Based solely on the possible atmospheric hazards of a confined space, it is easy to see why an unprepared rescuer can become a victim. Firefighters do not start fire operations without sizing up the situation, and EMTs do not start treatment on their patients without assessing them. Why would a rescuer at a confined-space accident want to do any less? A good analogy for confined-space rescue is to consider how you would rescue a person who was drowning. Would you jump into the water and attempt to swim to the victim, or would you call for help and then find other equipment such as a boat or a rope to help him? Well, your victim is drowning in an atmosphere that is toxic, oxygen deficient, or explosive. What equipment do you need to complete a safe and successful rescue?

Physical Hazards

▶ **physical hazards**

hazards within a confined space which are produced by mechanical, electrical, chemical, or thermal means and endanger personnel in the confined space.

Not all hazards faced in a confined space will be atmospheric hazards. **Physical hazards** must also be considered, including electrical equipment, agitators or paddles (shown in Figure 1–7), fire suppression equipment, sharp edges (shown in Figure 1–8), difficult-to-access areas due to height, and the physical state of the material present in the space. A common physical hazard presented by the materials is engulfment. A liquid or finely divided (flowable) solid substance can surround and trap a person and as a result, when the material is aspirated, can cause death by filling or plugging the respiratory system. At times the material can exert enough force on the body to cause death by strangulation, constriction, or crushing. Do not overlook physical hazards as you characterize the problems presented by the confined-space rescue.

FIGURE

There can be a variety of physical hazards present in a confined space. Note the sludge thickener (screw) running up the center of the picture.

FIGURE

This is a final clarifier tank in a sewage treatment plant. The hazards shown here include the water, sloped sides near the top of the tank, and the weirs (pointed metal triangles).

▶ **Permit-required confined spaces**

confined spaces that meet the definition of a confined space and have one or more of these characteristics: (1) contain or have the potential to contain a hazardous atmosphere, (2) contain a material that has the potential for engulfing an entrant, (3) have an internal configuration that might cause an entrant to be trapped or asphyxiated by inwardly converging walls or by a floor that slopes downward and tapers to a smaller cross section, and (4) contain any other recognized serious safety or health hazards.

NON-PERMIT CONFINED SPACES VERSUS PERMIT-REQUIRED CONFINED SPACES

To better differentiate between confined spaces that are expected to present hazards to people who enter them and those that are not, OSHA has classified confined spaces as **permit-required confined spaces** and **non-permit confined spaces.** A non-permit confined space is an area *which does not contain or, with respect to atmospheric hazards, have the potential to contain any hazard capable of causing death or serious physical harm.*

A permit-required confined space is an area that has one or more of the following characteristics:

1. Contains or has the potential to contain a hazardous atmosphere;

▶ **Non-permit confined spaces**

confined spaces that do not contain or, with respect to atmospheric hazards, have the potential to contain any hazard capable of causing death or serious physical harm.

NOTE: Rescuers must be aware of the difference between permit-required and non-permit confined spaces because either will affect the types of precautions taken to protect workers in the confined space.

NOTE: It is in your best interest to treat all confined spaces as permit-required spaces.

2. Contains a material that has the potential for engulfing an entrant;

3. Has an internal configuration such that an entrant could be trapped or asphyxiated by inwardly converging walls or by a floor which slopes downward and tapers to a smaller cross section; or

4. Contains any other recognized serious safety or health hazard.

Rescuers must be aware of the difference between permit-required and non-permit confined spaces because either will affect the types of precautions taken to protect workers in the confined space. The requirements to protect workers in the permit-required space may include ventilation, respiratory protective equipment, personal protective clothing, monitoring equipment, and retrieval equipment. A non-permit space may have none of this equipment present, and you may be tempted to consider it not as a true confined-space rescue; however, simply because a confined space was initially defined as a non-permit space does not mean that it contains no hazards. The work being performed in the space may have introduced new or unknown hazards into it. These new hazards may easily turn that non-permit space into one with serious hazards—so serious, in fact, that the space would be reclassified as a permit-required space. Further complicating matters is the fact that the OSHA standard does not apply to agriculture, construction, or shipyard employment. Any of these types of sites may have confined spaces present, but you will have to identify them as such and take the proper precautions. It is in your best interest to treat all confined spaces as permit-required spaces. Only after the space has been classified by monitoring, assessing, and reviewing the problem should you consider downgrading it to a non-permit space. If you have any doubt as to the need for a permit, treat the space as a permit-required confined space.

CASE ■ STUDY

It is a warm summer day, and you have been called to a construction site where a sewage pumping station is being built. Your initial report is that a construction worker fell into an excavation. The excavation is 15 feet square and the depth is 27 feet, as shown in Figure 1–9. The ground where the pumping station is being built is marshy and has a high water table. The pit is shored with vertical steel sheeting which extends above the top of the excavation. An unsecured ladder leads to an intermediate ledge (called a wale), and a second unsecured ladder leads from the wale to the bottom of the pit. The site includes pumps that are used to pump water from the pit, but they are not running at this time. As you look into the hole, you see a construction worker lying at the base of the excavation. You also see water and sand "boiling" from a partially driven piling at the bottom of the hole and you detect a smell of rotten egg.

FIGURE

Elevation view of the confined space case study in the text.

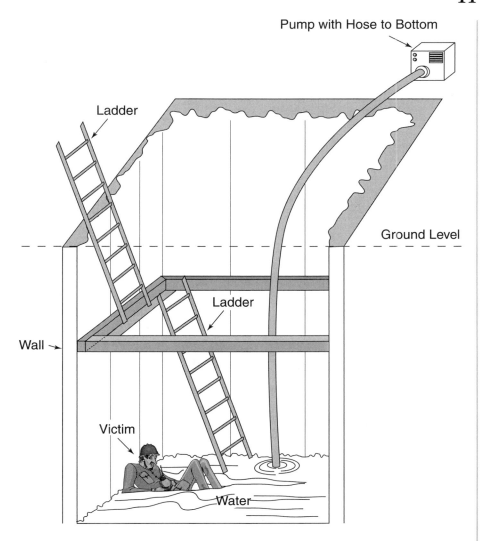

Questions to Consider

1. Does this construction pit meet the definition of a confined space?
2. What hazards do you know or suspect to be present?
3. What can you do to prevent further injury?

Answers

1. This construction pit meets the definition of a confined space in that it (1) has limited openings for entry and exit; (2) has unfavorable natural ventilation, which can lead to the presence of an atmosphere that is immediately hazardous to life and health; and (3) is not designed for continuous occupancy.

2. The "rotten egg" smell is an indicator of **hydrogen sulfide,** an extremely toxic, colorless, flammable gas more often called swamp gas or marsh gas, and pockets of it are common in marshy soil. It is produced by the decomposition of organic material. In addition, hydrogen sulfide can cause a temporary

▶ **hydrogen sulfide**

a flammable, toxic gas that can be created by the decomposition of organic materials.

▶ **methane**

a colorless, odorless, flammable gas consisting of one carbon atom and four hydrogen atoms, also called natural gas.

▶ **vapor density**

the weight of a given volume of gas or vapor compared to an equal volume of air at the same temperature and pressure. Air has a vapor density of 1. Gases with vapor densities less than 1 are lighter than air, whereas vapor densities greater than 1 are heavier than air.

loss of the sense of smell, which can cause people to believe that the gas is no longer present. Although you should not smell for the presence of a hazardous atmosphere, the smell of gas is often the first clue to its presence.

The fact that the water and sand at the bottom of the pit appear to be boiling indicates that gas is rapidly escaping from the ground into the pit. You should first ponder the source of this gas. Is it possible that an underground pipe has been ruptured during construction? Is so much gas escaping from the ground that the gas is bubbling through the water so rapidly that it appears to be boiling? In this particular incident, the source of the gas was only that which was present in the ground.

Besides the presence of hydrogen sulfide, you should also suspect **methane** in this type of soil due to the decomposition of organic material. Methane is an odorless, colorless, flammable gas. Even though most of us know methane as natural gas and can detect a characteristic odor, "pure" methane has no odor. (The familiar smell from natural gas is actually another chemical added to the gas.) The potential for a flammable atmosphere exists.

The oxygen level is also unknown in this space. Because hydrogen sulfide is heavier than air, it easily can displace the air at the bottom of the pit. The relative weight of a given volume of gas is referred to as its **vapor density.** The vapor density of air is given as 1, and all other gases are compared with air. Hydrogen sulfide has a vapor density of 1.2 and is thus heavier than air. Methane has a vapor density of 0.55 and is thus lighter than air. If this confined space were covered, there would be a layer of hydrogen sulfide at the bottom, a layer of air in the center, and a layer of methane at the top, as shown in Figure 1–10.

During the incident, a number of other people were reported to have inhaled hydrogen sulfide while standing around the excavation. Besides the atmospheric hazards present, the physical hazards are the unsecured ladders, the depth of the pit, and the water in the bottom of the pit. You may want to use the ladders already in place, but first secure them. You will also need to provide fall protection when entering the pit, which will double as a safety line to let you retrieve your own people in the event that they are injured during entry. The water presents another problem in that it is flooding the hole, and the pumps are shut down. Consider why this is so, which in this particular case is because the pumps were shut down to allow the bottom of the excavation to fill with water to hold it from blowing out.

Other physical hazards may include electricity, unsecured equipment both around the opening of the excavation and in it, and construction equipment that is being operated by construction workers as they attempt to either rescue or assist in the rescue of their fellow construction worker.

FIGURE 1-10

The vapor density of materials varies from material to material. Air has a vapor density of 1. Materials with a vapor density greater than 1 are heavier than air, while those with a vapor density less than 1 are lighter than air.

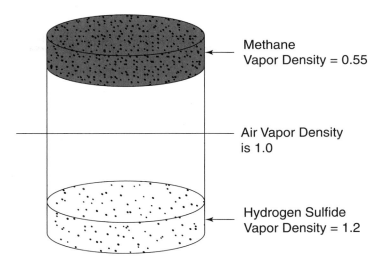

Methane
Vapor Density = 0.55

Air Vapor Density
is 1.0

Hydrogen Sulfide
Vapor Density = 1.2

3. First, you must recognize that this is a confined-space rescue operation. To prevent further injury, you must take control of the incident and secure the scene. You must not allow entry into the space by anyone who is not equipped and trained for confined-space rescue. You must begin assessing the scene and start monitoring the air both in the space and in the area surrounding it to determine the extent of the hazardous area. Look beyond the victim and try to determine what has happened, what is happening now, and what is expected to happen as the incident progresses.

In this particular incident, one police officer was killed attempting to rescue the construction worker. The ironic part is that there was a similar incident the day before during which a construction worker fell into the pit, was severely injured, but was successfully rescued. The initial call for the second emergency was that a man "had collapsed in the hole again."

■ SUMMARY

Confined-space emergencies have the potential to kill workers and rescuers. Only by recognizing that the emergency is a confined-space emergency and knowing of the dangers that can occur therein can you hope to protect yourself as you attempt to rescue the victim. All emergency rescuers want to be successful but do not want to die or be seriously injured in the process. Recognizing that a confined space is large enough to enter, has limited means for entry or exit, and is not designed for continuous human occupancy is only the beginning of a confined-space rescue operation. You must also realize the potential atmospheric and physical hazards present in the space, and the effects they will have on your ability to operate there. Any of the operational problems presented by a confined space can dramatically change your operations and may require you to call for additional help. Professional rescuers recognize the existence of such problems and accept that it is part of the cost of doing business. Maintain the costs in terms of human life as low as possible.

■ REVIEW QUESTIONS

1. OSHA defines a confined space by identifying what three separate items that characterize the space?

2. If you respond to an emergency that has occurred in what appears to be a confined space, but only one of the conditions established by OSHA is present, should you treat the space as a true confined space?

3. The flammable range of a material has a lower flammable limit and an upper flammable limit. Are the flammable range and lower and upper limits identical for all materials?

4. What is the range for acceptable oxygen concentrations within a confined space?

5. Identify two potential physical hazards within a confined space.

6. The vapor density of air is given as 1. Is a gas which has a vapor density of 3.1 lighter or heavier than air?

7. Using the sense of smell to detect the presence of hazardous gases or vapors is a recommended practice. True or false?

8. For combustion to occur, there must be _____.
   a. fuel, heat, and oxygen
   b. sufficient fuel, heat, and oxygen
   c. fuel, heat, and an uninhibited chemical reaction
   d. heat, an uninhibited chemical reaction, and oxygen

9. The process by which a liquid or finely divided solid substance surrounds and traps a person is known as _____.
   a. asphyxiation
   b. strangulation
   c. engulfment
   d. physical hazard

10. The flammable range for gasoline is _____ and the flammable range for acetylene is _____.
   a. 1.4 to 7.6 percent in air, 15.5 to 27 percent in air
   b. 2.5 to 100 percent in air, 1.4 to 7.6 percent in air
   c. 2.5 to 100 percent in air, 15.5 to 27 percent in air
   d. 1.4 to 7.6 percent in air, 2.5 to 100 percent in air

C H A P T E R

2 Confined-Space Entry Requirements

OBJECTIVES

After completing this chapter, the reader should be able to identify the functions of the following positions within a confined-space program:

- attendant
- authorized entrant
- confined-space supervisor

and define the following terms:

- permit-required confined space
- non-permit confined space
- confined-space permit

INTRODUCTION

In reviewing the information contained in Chapter 1, it becomes obvious that most if not all confined-space accidents can be prevented. In 1993 the U.S. Department of Labor through OSHA adopted regulations that required employers to develop a program to prevent confined-space accidents. Even before the adoption of the federal regulations, however, some states and many private companies developed programs for confined-space entry. Although some of these programs were simple and others were more detailed, their goal was still the same: to provide for worker safety during entry into confined spaces. An interesting example of such a program was used in the beer brewing industry. After the brewing tanks were emptied, workers had to enter and clean them. In consideration that the beer was meant for consumption, the only atmospheric hazard that remained in the tank after brewing was the carbon dioxide. The tanks would be ventilated and then tested for oxygen content by lowering a miner's lantern into the tank. If the lantern continued to burn, the oxygen level was adequate to support human life; if the lantern extinguished, then the atmosphere was oxygen deficient. This procedure may sound primitive, but it was an effective method until other means became available to test the atmosphere in the tank. The problem, however, was that only some companies had confined-space programs. This discrepancy within industry led to the development of the OSHA Confined-Space Standard, CFR 1910.146, described in Chapter 1.

In addition to providing for safe entry into confined spaces, the OSHA regulations require a means to rescue people from the confined space in the event of an accident. This requirement is where your involvement as a rescuer begins—with a need for safe entry of workers into the space (as shown in Figure 2–1) and of rescuers attempting to help people who might have been injured or trapped by an accident. If an accident occurred in the confined space, yet the purpose of a confined-space program is to prevent an accident, then something went wrong. This "something" can range from the absence of a confined-space program by the employer to the accidental introduction of a hazard into the space after the program was initiated. You must be familiar with the requirements of a confined-space program to determine what has caused the emergency. You must also be familiar with the requirements of a confined-space entry program so that your emergency organization can develop standard operating procedures (SOPs) to protect you—the rescuers—during an emergency.

As you view the requirements for confined-space entry, it should become obvious that most of the requirements for a program are valuable for sizing up the incident. The **entry permit** is a written document which outlines what will occur within the space. If no permit is present, you will require more time to determine what has happened and what steps you must take to protect yourself. Even with a permit present, you will need to verify that the information is

▶ **Entry permit**

a written document which must be completed prior to entry into a confined space and defines the hazards of the space, the precautions to be taken, the type of work that will be performed, the roles of personnel involved in the entry, and other specific details.

correct. Knowing that the entry program requires people to be trained as attendants, entrants, or supervisor will enable you to begin accounting for the workers at the scene. Being aware that someone is missing could indicate that there are more victims than first believed. Over 60 percent of the fatalities in confined-space accidents are untrained rescuers. There might be more than one victim in the confined space.

| NOTE: Over 60 percent of the fatalities in confined-space accidents are untrained rescuers.

CONFINED-SPACE PROGRAMS

For an employer, a confined-space program begins with evaluating the workplace to determine if any permit-required confined spaces exist. If the workplace contains such spaces, the employer must use signs or other effective measures to inform employees who might be exposed to the hazards of the space. These signs may say "Danger—Permit-Required Confined Space, Do Not Enter" or use other language to effectively communicate the presence of a permit-required confined space (see Figure 2–2). But what if the workplace survey only reveals non-permit confined spaces? Those signs will not be posted. Does that mean that those non-permit spaces will not pose a hazard? The answer is no. Rescuers should treat all confined spaces as potentially hazardous until a thorough evaluation of the conditions shows that no hazard exists.

SAFETY

Rescuers should treat all confined spaces as potentially hazardous until a thorough evaluation of the conditions shows that no hazard exists.

The employer may decide that her employees will not enter the confined spaces, but if you are called to make a rescue from a confined space, how did that person get into the confined space? Is the person an employee who entered the space in violation of company policy, or the employee of an outside contractor working at the site? Either way, you must take control of the situation.

The simplest way to begin to define a confined-space program is to identify the roles and responsibilities that people are given during such an entry. Although these roles were intended for workers entering the space to perform work, these same roles can easily be

FIGURE **2-2**

Confined spaces are required to be identified as part of a confined space entry program.

adapted for rescue purposes. To keep these definitions simple, the roles and responsibilities identified here will only be those required by OSHA for workers who will enter and work within the confined space. The roles and responsibilities for rescuers will parallel those of the worker, but, as you work your way through this book, the differences should become clear.

Attendant

The OSHA standard includes a definition and a requirement for an **attendant**. The definition of the attendant's functions is an individual who is stationed outside one or more permit-required confined spaces and monitors the people within and the conditions in the area around the space, as shown in Figure 2–3. The attendant is not to enter the space to bring tools or equipment into it, to perform rescue, or to enter for any other reason unless relieved by another trained attendant.

The attendant is also responsible for monitoring the people within the space to ensure their safety and to ensure and maintain communication with the personnel within the space. The attendant can call for help if an accident occurs within the space and can order its evacuation in the event that conditions either within or outside of the space pose a hazard to any person therein.

The attendant's role is critical because that person is the outside link for those in the confined space. The attendant not only monitors conditions inside but also watches for hazards outside. With no attendant, how would the workers in the space know that a fire alarm was sounding in the building where the confined space was located? With no attendant, who would be responsible for ensuring that the area around the opening to the space was clear of tools and equipment, as shown in Figure 2–4? These tools could easily be kicked or knocked into the space or onto any workers there. During any potential rescue situation, the attendant is the single most valuable source of information for rescuers. The attendant must remain alert to the situation and avoid feelings of responsibility for the accident so as to not give out inaccurate information.

► **Attendant**

the person trained and assigned to remain outside of the confined space, monitor conditions inside and outside of the space, and communicate with persons inside.

SAFETY
The attendant is not to enter the space to bring tools or equipment into it, to perform rescue, or to enter for any other reason unless relieved by another trained attendant.

NOTE: The attendant's role is critical because that person is the outside link for those in the confined space.

FIGURE **2-3**

Attendant communicating with the entrants working inside the confined space.

FIGURE **2-4**

Tools and equipment lying around the outside of a confined space present a danger in that they can fall or be kicked into the space.

▶ **Authorized entrants**

those persons trained, assigned, and equipped to enter and work within the confined space. During a confined-space rescue, those persons trained, equipped, and assigned to enter the confined space are known as the rescue entrants.

Authorized Entrant

The purpose of any confined-space program is either to make the area safe for people to enter or to keep everyone out. As shown in Figure 2–5, the people who will enter the space to work have roles and responsibilities. When they are made aware of these requirements, they will become **authorized entrants**. Basically, the authorized entrants must be familiar with the hazards of the confined space and must know how to use required personal protective equipment, recognize any warning signs or symptoms of a dangerous condition, alert the attendant as to when a dangerous condition is detected, and exit from the space as quickly as possible when necessary.

The entrant is the person who probably will be the victim during an emergency at a confined-space accident. Not all entrants are authorized to be in the space. Some victims may be those who attempted to enter without training or who attempted the rescue of another

FIGURE 2-5

Any person who makes entry into a confined space is an entrant and must be trained as an entrant. This entrant training requirement includes rescuers.

person in the space. Obviously it takes a good amount of training for a person to qualify as an authorized entrant. If, after completing the training, the entrant still became a victim, then an immediate investigation of the situation would be required.

Confined-Space Supervisor

Considering the requirements of a confined-space entry program, a person must be placed in charge of the program during permit-required entries. This job falls to the person OSHA identifies as the **confined-space supervisor.** This person has a variety of tasks and obligations to fulfill. The supervisor is required to:

▶ **Confined-space supervisor**

the person assigned responsibility for ensuring that the requirements of a confined-space program have been met prior to, during, and after entry by persons into a confined space.

- ■ know the hazards that may be faced during entry, including the mode, signs or symptoms, and consequences of the exposure and inform the attendant and entrant of the hazards;

- ■ verify, by checking that the permit has been filled out properly, that all tests specified by the permit have been conducted, and that all procedures and equipment required by the permit are in place before allowing entry to begin, as shown in Figure 2–6;

- ■ terminate the entry and cancel the permit as required;

- ■ verify that rescue services are available and that there is a means for summoning them;

- ■ remove unauthorized individuals who enter or who attempt to enter the permit space; and

- ■ determine that entry operations remain consistent with terms of the entry permit and that acceptable entry conditions are maintained at all times.

In effect, the entry supervisor should be the most knowledgeable person on the scene of a confined-space accident, but that supposition is based on the existence of a confined-space entry program.

FIGURE

The confined-space supervisor is responsible for making sure that the entry permit is properly filled out.

As discussed earlier in this chapter, the intent of such a program is to prevent accidents, but if an accident occurs, who or what is the cause? Perhaps a hazard introduced into the space caught the workers off guard, or perhaps no confined-space entry program was initiated. Without a program there will be no entry supervisor, no trained entrants, and no trained attendant. In short, there will be very little accurate information that you can develop from trained people at the scene. You may have to begin at the most basic level to protect yourself or other emergency responders. In addition to preventing an accident a confined-space program can minimize the effects of an accident and speed your rescue operations. The information you will need to make decisions during a rescue will only be as good as the source of that information.

┃ NOTE: A confined-space program can minimize the effects of an accident and speed your rescue operations.

Confined-Space Entry Permit

In consideration of the numerous items that must be reviewed before a permit-required confined-space entry, personnel use a standardized checklist at the confined space, as shown in Figure 2–7, which is basically the confined-space entry permit. To help ensure that the hazards (both atmospheric and physical) are reviewed prior to an entry, certain requirements must be met for a basic confined-space entry permit. These items range from the simple to the technical, but the problems are no less significant for the simple items, such as correctly identifying the confined space, than they are for the more complicated items, such as providing chemical protective clothing. In fact, it is easy to miss the simple problems because rescuers take them for granted.

┃ NOTE: To help ensure that the hazards (both atmospheric and physical) are reviewed prior to an entry, certain requirements must be met for a basic confined-space entry permit.

A confined-space entry permit which meets the intent of the OSHA standard should address the following information:

■ Date and time the permit was issued and the date and time the permit expires. No permit should be issued indefinitely.

ENTRY PERMIT

PERMIT VALID FOR 8 HOURS ONLY, ALL COPIES OF PERMIT WILL REMAIN AT JOB SITE UNTIL JOB IS COMPLETED

DATE - - SITE LOCATION and DESCRIPTION_____

PURPOSE OF ENTRY_____

SUPERVISOR(S) in charge of crews	Type of Crew	Phone #

COMMUNICATION PROCEDURES _____

RESCUE PROCEDURES (PHONE NUMBERS AT BOTTOM)_____

BOLD DENOTES MINIMUM REQUIREMENTS TO BE COMPLETED AND REVIEWED PRIOR TO ENTRY

REQUIREMENTS COMPLETED	DATE	TIME
Lock Out/De-energize/Try-out	____	____
Lines(s) Broken-Capped-Blanked	____	____
Purge-Flush and Vent	____	____
Ventilation	____	____
Secure Area (Post and Flag)	____	____
Breathing Apparatus	____	____
Resuscitator - Inhalator	____	____
Standby Safety Personnel	____	____
Full Body Harness w/"D" ring	____	____
Emergency Escape Retrieval Equip	____	____
Lifelines	____	____
Fire Extinguishers	____	____
Lighting (Explosive Proof)	____	____
Protective Clothing	____	____
Respirator(s) (Air Purifying)	____	____
Burning and Welding Permit	____	____

Note: Items that do not apply enter N/A in the blank.

**RECORD CONTINUOUS MONITORING RESULTS EVERY 2 HOURS

CONTINUOUS MONITORING**	Permissible	
TEST(S) TO BE TAKEN	Entry Level	
PERCENT OF OXYGEN	19.5% TO 23.5%	
LOWER FLAMMABLE LIMIT	Under 10%	_____
CARBON MONOXIDE	+35 PPM	_____
Aromatic Hydrocarbon	+1 PPM*5PPM	_____
Hydrogen Cyanide	(Skin)*4PPM	_____
Hydrogen Sulfide	+10PPM*15PPM	_____
Sulfur Dioxide	+2PPM*5PPM	_____
Ammonia	*35PPM	_____

*Short-term exposure limit: Employee can work in the area up to 15 minutes

+8 hr. Time Weighted Avg. Employee can work in area 8 hrs (longer with appropriate respiratory protection).

REMARKS: _____

GAS TESTER NAME & CHECK	INSTRUMENT(S) USED	MODEL &/OR TYPE	SERIAL &/OR UNIT #
_____	_____	_____	_____
_____	_____	_____	_____

SAFETY STANDBY PERSON IS REQUIRED FOR ALL CONFINED SPACE WORK

SAFETY STANDBY PERSON(S)	CHECK #	CONFINED SPACE ENTRANT(S)	CHECK #	CONFINED SPACE ENTRANT(S)	CHECK #
_____	_____	_____	_____	_____	_____
_____	_____	_____	_____	_____	_____

SUPERVISOR AUTHORIZING   ALL CONDITIONS SATISFIED _____

DEPARTMENT/PHONE _____

AMBULANCE 2800   FIRE 2900   SAFETY 4901   GAS COORDINATOR 4529/5387

FIGURE 2-7

This is a sample confined-space entry permit.

Conditions both inside and outside the space change over time. A noncurrent permit is an invitation to disaster, because it gives workers a false sense of security with the result being that they are not working in a safe environment. You should also be aware that some permits are issued for as long as a year, but expected hazards in these spaces are very low.

■ Correct identification of both the job site/space and the supervisor. Employees must know that they are entering the correct space and that they have properly identified the hazards of the space they are entering. The supervisor must also be identified so that the workers can get answers to any questions that they may have about the space, including the signs and symptoms of exposure and the potential hazards.

■ The equipment to be worked on and the type of work to be performed. It is important to make the space safe from unexpected and deadly hazards, such as accidental start-ups of equipment, the accidental opening of a valve, welding inside a confined space, and so on. By knowing what type of work is to be performed and where it is to be done, the workers can protect themselves and others from injury or death.

■ Correct identification of the attendant and the entrants. In this way all personnel know who belongs at the job site and what is expected of them. If a person is identified as the attendant, she is expected to be present and perform all duties of the attendant. The entrant and attendant have different duties and each must be trained to perform those duties. Providing proper identification is paramount to knowing when the proper people are present and doing their job. In addition, using this procedure provides a head count of the personnel working in and near the space, so the attendant and the supervisor know who is expected to be present at the site.

■ Atmospheric checks before and during entry. How could one person take all of the readings and remember them without some type of checklist? The entry permit requires that readings be taken for oxygen levels, flammable gases (also called explosive gases), and toxic gases based on the known or suspected hazards in the space. Additionally, the tester must sign the permit to confirm that the readings have been taken and recorded.

► **Lockout/tagout**

elimination and control of hazardous sources of energy or products.

► **Blinding**

the insertion between two flanges of a device called a blind which has no opening in it and is meant to prevent the flow of a product past the blind.

■ The elimination and securing of sources of potential physical or chemical hazards to isolate them from the space. This procedure is called **lockout/tagout** and will be discussed in greater detail in chapter 4. Included in these types of lockout/tagout procedures are items such as shutting down pumps, capping product lines that enter the confined space (also called **blinding**), and disconnecting or blocking the lines.

24

▶ **Permissible exposure limit
(PEL)**

a time weighted average (TWA)
concentration that must not be
exceeded during any 8-hour
work shift of a 40-hour work
week. A short-term exposure
limit (STEL) is measured over a
15-minute period.

| NOTE: If your group or
agency is the designated
rescue team, then you have
the right to inspect the
confined space for
preplanning a rescue.

■ A thorough check for atmospheric hazards. Atmospheric hazards present the greatest risk to people in a confined space. To minimize those hazards it may be necessary to provide mechanical ventilation for the space. Natural ventilation is acceptable where conditions permit, but if the permit indicates that mechanical ventilation is required then it must be in place.

■ A thorough check to ensure that the ventilation and isolation have been effective after the confined space has been isolated and vented. To accomplish this task, someone must take meter readings for oxygen, combustible gases, and toxic gases. The acceptable levels of each atmospheric hazard should be identified on the permit. For oxygen, a level between 19.5 and 23.5 percent in air is required. For combustible gases the level must be less than 10 percent of the lower flammable limit. For toxic gases the level must not be greater than the **permissible exposure limit (PEL)** of the toxic material in air.

■ Communication procedures outlined on the permit so that the attendant can check with the entrants. This responsibility is especially important if the entrant will be out of the attendant's sight. The attendant must not enter the confined space for any reason unless another trained person takes the role of attendant. Effective communication procedures protect both the attendant and entrants.

■ Rescue procedures outlined on the entry permit, which is the beginning of your involvement. If your group or agency is the designated rescue team, then you have the right to inspect the confined space for preplanning a rescue. When your agency or group is not the designated rescue team, you may come into a complicated situation that is beyond the ability of the on-site rescue team, or you might be called in for a support role for the rescue. Knowing in advance about the types of on-site confined spaces and the hazards they present will save time and lives—maybe yours. Cooperation and preplanning are keys to an efficient operation in this case.

■ Identification of all entry, standby, and backup persons on the permit. This procedure is designed to keep unqualified people out of the space. Not only must the entry, standby, and backup persons be identified, but they must also have successfully completed all required training and it must be current. This example is another reason why a permit becomes so valuable for the entry. Imagine if you were the entry supervisor and had to remember all of the items without a checklist. Do you think that you could do it accurately enough to protect another person's life?

Not only does a permit-required entry call for different types of equipment, but the equipment must also be at the worksite. A checklist prompts the employees and supervisors to avoid taking shortcuts. The equipment should also be functioning properly. What good is a combustible gas detector if it is not accurate or does not work? What about a defective self-contained breathing apparatus (SCBA)? Not having functional protective equipment at a confined-space entry worksite can actually be the cause of the accident when all other precautions have been taken. If it is worth doing, then it is certainly worth doing correctly. Among the equipment that needs to be considered for use at a confined-space entry are the following:

- direct-reading gas monitors, as shown in Figure 2–8
- safety harnesses and lifelines for entry and standby persons, as shown in Figure 2–9

FIGURE 2-8

Shown here are several different types of direct reading monitoring devices. Included are a combustible gas detector (meter on right) and several multiple gas detectors.

FIGURE 2-9

Harnesses allow a rescuer to be lowered into a confined space and retrieved should they become incapacitated.

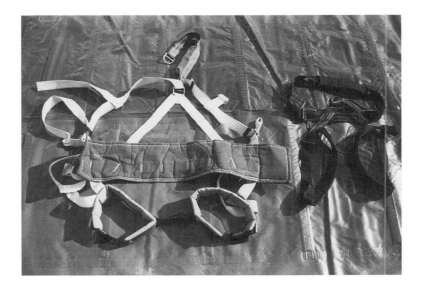

FIGURE 2-10

These are spare air cylinders specifically set up for a supplied air respirator. Having rescue and safety equipment which is designed for use at a confined-space rescue improves rescuer safety and enhances operations.

▶ **Class I, Division I, Group D**

electrical equipment specified under the National Electrical Code as meeting particular requirements for safe performance under certain conditions. The designation of class, division, and group refers to distinct hazardous atmospheres that may be present during the use and operation of this equipment.

NOTE: If the confined-space atmosphere is monitored, readings are below 10 percent of the LEL, and ventilation is maintained, then the nonsparking tools may not be needed. This decision should be made by the entry supervisor.

◐ **SAFETY**

Use only manually powered hoists for raising and lowering people in and out of a confined space.

- hoisting equipment
- communications
- SCBAs for entry and standby persons
- protective clothing, as shown in Figure 2–10
- electric equipment listed **Class I, Division I, Group D**
- nonsparking tools

If the confined-space atmosphere is monitored, readings are below 10 percent of the LEL, and ventilation is maintained, then the nonsparking tools may not be needed. This decision should be made by the entry supervisor.

Some of the equipment listed here is intended to prevent an accident in the confined space. Nonsparking tools will not be an ignition source nor would the listed electrical equipment. Other considerations should include electrocution hazards presented by extension cords and electrical tools, the need for chemical protective clothing, hoisting equipment that is stable and designed to be used to hoist people in and out of the space, and hoisting equipment for tools and materials that must be raised or lowered into the confined space. One word of caution about hoisting equipment: Use only manually powered hoists for raising and lowering people in and out of a confined space. Electrically driven or mechanically driven equipment operates at too fast a speed and provides too much force to keep from injuring a person who may be caught by something within the space while being hoisted. Manually powered equipment is usually slow enough and does not produce so large a force as to easily injure the person being hoisted. Periodic atmospheric tests are required to maintain a safe confined-space entry. The types of monitoring and the frequency will vary with the space and the expected hazards.

The entry supervisor or a person in charge must sign the permit. Not all entrants are required by OSHA regulations to sign the permit; however, all entrants should confirm that the permit has been filled out and that conditions are acceptable to enter. Depending on the internal requirements within certain companies, there may be a space at the end of the permit for the employees who will be entrants and/or attendants to sign. This signature is intended for internal use only. All employees should receive written instructions and safety procedures which they must understand. No one should be expected to enter the confined space if any of the questions on the permit are answered as no. All spaces on the permit should be marked either yes, no, or not applicable. Any time the question is answered as no, it is intended to mean that the hazard or condition exists and has not been properly addressed. No entry permit is valid unless all appropriate items required by the permit are completed. Employees should not treat this section as an administrative formality. This section is intended to remind them of their rights to work in a safe manner. Other signatures at the end of the permit identify the person who prepared the permit (the entry supervisor), and the unit supervisor in whose area the confined space is found.

A copy of the permit must be kept at the job site, as shown in Figure 2–11. As an emergency responder, this becomes a valuable source of information to you. The permit is required to be on-site as long as entry operations by employees are ongoing. In the event that the permit is not at the job site because the original work was completed, a copy of the permit must be kept at the facility for a year after the entry. If you are called to a confined-space accident in which the work was completed but someone made entry without a valid permit, then the original permit may help you in evaluating the situation.

CONFINED-SPACE PROGRAM AND RESCUERS

When looking at a confined-space entry program it is easy to believe that it is too complicated. The reality is that it is not difficult, but rather new to many rescuers. You will find that by understanding the roles of the attendant, entrant, and supervisor you will identify key roles that rescuers must fill to provide for the safety of the rescue team. You might not call the rescuers who will enter the confined space "entrants," but that is what they are. The same level of knowledge about the confined space and its hazards is required by the rescue entrant. Your rescue attendant might be called a "safety officer" or some other name instead of attendant, but what tasks does she perform which are different than the attendant's when that person is assigned to monitor the confined space and the people who enter the space during a rescue? The role of the entry supervisor might actually be filled by the person who is the incident commander during simple rescue operations. At other times the entry supervisor might also have to be the safety officer. The rescue attendant's role would change to monitor the rescuers in the space. The point here is that the title of the person performing the task is not as important as identifying the needed tasks and assigning people to perform them. Understanding a confined-space entry program and adapting it for rescue purposes protects your rescue team members and you. Simplify your

FIGURE 2-11

FIGURE 2-11

A copy of the confined-space entry permit must be kept at the job site. This particular permit is a long-term permit (issued annually) and spells out what work is to be done and who is to do it.

ANNUAL CONFINED SPACE ENTRY PERMIT

Location: __Primary Pump Station - Dry Well__   Date Posted: __June 19, 1999__

Permit Expires: __June 18, 2000__

Description of Work Normally Performed: __Routine Operator check, maintenance and cleaning.__

_____

_____

Persons Authorized to Enter (Positions): __Operations and Maintenance Departments personnel__

__and authorized visitors with escorts.__

Atmospheric Testing (Perform Prior to Posting)

Time: _____0846_____   Oxygen Level: _____20.9_____ %

Make of Unit: __CGM__   Toxic Level: _____0_____ PPM

Serial No: __9161__   Combustible Level: _____0_____ %

Atmospheric Testing Performed by: __George Browne/__ _George Browne_
                                                    Print/Signature

Above Results Confirmed by: __Gus S. Crist/__ _Gus S. Crist_
                                                    Print/Signature

Any special instruction for this entry?:   Yes: __X__   No: _____
(i.e., gas detector, ventilation, etc.)

Note special instructions here: __When ventilation is on, no gas detector is required. When ventila-__

__tion is not operational, a gas detector is required for entry.__

Note: This Annual Permit will be revoked whenever any conditions in the confined space have become more hazardous than contemplated.

Supervisor Authorizing this Annual Permit: __Gregory Hansson/__ _Gregory Hansson_
                                                    Print/Signature

Gregory Hansson, Director, Eastern Division

h:\ops\confspps.xls   7/99

---

need to identify what steps you must take to ensure your safety, the safety of the victim, and the equipment you will need to perform the rescue operations. Look at the requirements of the OSHA confined-space entry program. Adapt those requirements to fit your needs.

Permit-Required versus Non-Permit Confined Spaces

In Chapter 1 we discussed permit-required and non-permit confined spaces. Clearly, a permit-required confined space must have an entry permit present, but what about a non-permit confined space? As

FIGURE 2-12

Not all spaces are permit-required confined spaces. This space does not require a confined-space entry permit since it does not meet the definition of a permit-required confined space. But if an emergency occurred, how would you handle this space?

| NOTE: Just because the space was originally a non-permit required space does not mean that you should ignore the precautions laid out in the confined-space entry program.

long as the non-permit space is expected to be free of hazards, you will not find a confined-space permit at the worksite, as shown in Figure 2–12. A word of caution: Just because the space was originally a non-permit required space does not mean that you should ignore the precautions laid out in the confined-space entry program. Non-permit spaces may have work being performed in them which introduces a hazard there. Something as simple as painting the inside of a non-permit space can lead to a lowered oxygen level or the presence of toxic vapors. Your own awareness and caution should dictate that at any confined-space accident you treat the space as permit required, use some basic precautions, and determine if the area is free of hazards by using monitoring equipment and assessing the physical hazards.

■ SUMMARY

Prevention of injuries and death is the purpose of a confined-space entry program. If you have been called to a confined-space accident, something has gone wrong at the site. Even when a program is in place and an accident has occurred, there are still benefits that emergency workers can take from it. A proper permit will assist you in determining the potential hazards of the confined space. Even though something is wrong that was not identified on the permit, it still provides a starting place. Additionally, there should be an attendant, supervisor, and possibly an entrant at the scene. Take the time to interview these people to find out what has happened. When a company has no confined-space program for the workers and an emergency occurs, you will be at a disadvantage and will have to develop the information on your own. Either way, understanding the basis of such a program should make you aware of the need to structure your rescue operations in a similar fashion. Chapter 9 will detail how to implement a confined-space rescue program based on a real program. The roles of attendant, entrant, and supervisor will change only slightly for rescue.

Where a confined-space entry program is in place, the area will have permit-required and non-permit confined spaces. For rescue operations, all confined-space emergencies should be considered as being in the same category as permit-required spaces. Only after you have monitored and investigated the conditions within the space, and it has been shown to be safe enough to downgrade to a non-permit level, should you change the classification of it.

■ REVIEW QUESTIONS

1. The role of the attendant is to stay outside of the confined space and monitor conditions both inside and outside the space. Explain the value of monitoring such conditions.

2. The attendant is not to enter the confined space for any reason. Why is this requirement so critical to the safety of people within the space?

3. Briefly explain how the roles of attendant, authorized entrant, and entry supervisor would be paralleled by the members of a rescue team.

4. Identify four pieces of information that are required to be contained on a confined-space entry permit and how the information would be valuable to a rescue team.

5. Explain the difference between a permit-required confined space and a non-permit confined space. How would this difference affect your rescue operations?

6. What is the main reason for an entry permit?
   a. To identify the number of personnel working in the confined space.
   b. It is a written document which outlines what is going on in the space.
   c. It is a sign-off sheet that everything is safe before entering.
   d. All of the above

7. The attendant is authorized to bring tools into the confined space when performing the attendant duties. True or False

8. To isolate a physical or chemical hazard you need to ____ ____ the area in question.

9. You should consider using an electrical or hydraulic hoist to raise or lower people in a confined space. True or False

10. Who is in charge of a confined-space entry?
    a. Entrant or entrants
    b. Attendant
    c. Confined-space supervisor
    d. All of the above

3 Air Monitoring

OBJECTIVES

After completing this chapter, the reader should be able to explain the purpose of the following monitoring equipment:

- combustible gas indicator
- oxygen meter
- specific gas monitoring equipment
- gas specific meter
- colorimetric tubes
- pH devices

and define the following terms:

- parts per million
- percentage of the lower flammable limit
- action limit
- pH scale
- calibration

INTRODUCTION

As you stand by a river swollen with flood waters, you hear a person who is trapped in the middle of the river cry for help. You look at the water, brown with mud, charging past with a force that is pushing along debris the size of a small car. Of what hazards are you immediately aware? What danger would stop you from attempting to rescue this person trapped in the river? Now imagine that you have been called to a chemical plant for a confined-space incident and are standing at the opening of a process vessel. Two workers lie unconscious at the bottom of the vessel and their coworkers are anxiously telling you to rescue their friends. There is no entrapment of the victims with any equipment within the space and the scene inside is quiet. At this incident, what hazards do you notice immediately? What dangers are present? Chances are that you will not see the atmospheric hazard. There is no way to visibly detect an atmospheric hazard and accurately determine the degree of danger it presents; hence, monitoring equipment plays an important role. By using the right monitoring equipment you can begin to characterize the type of hazard present and the degree of danger it presents to both the rescuers and the victims.

Numerous types of atmospheric monitoring equipment are available. For the purpose of confined-space rescue, this book discusses direct reading, real-time instruments only. As the name implies, **direct reading instruments** provide a direct reading either from a digital readout, a gauge of some type, a measurable change of color of a sample tube, or some other indicating method. The important fact is that the results are measurable. This measure can be in parts per million, percentage, pH, or other recognized measure. Certain monitoring devices merely detect the presence of a particular gas and do not give a direct reading (a home carbon monoxide detector is one example). The value of such instruments is limited in that the instruments can tell you that a specific gas is present, but not the quantity of that gas. In addition to providing a direct reading, the instruments give real-time readings in that the results are immediate. There is no time delay because a sample had to be sent to a lab for analysis. For emergency responders, real-time and direct reading instruments provide immediate, on-scene measurements, as shown in Figure 3-1.

By no means does this section attempt to be a complete guide to the use of monitoring instruments. You must consult the manufacturers' instructions and train with all monitoring instruments well in advance of an emergency. You must know not only how to use your instruments, but also how to maintain them. All monitoring instruments require periodic calibration and replacement of sensors. The manufacturer can provide you with needed guidelines.

COMBUSTIBLE GASES

To detect and measure the presence of combustible or flammable vapors, a **combustible gas indicator (CGI)** is used, as shown in

---

| **NOTE:** By using the right monitoring equipment you can begin to characterize the type of hazard present and the degree of danger it presents to both the rescuers and the victims.

▶ **direct reading instruments**
detection and monitoring instruments that give a reading based on a graduated scale.

| **NOTE:** For emergency responders, real-time and direct reading instruments provide immediate, on-scene measurements.

▶ **combustible gas indicator**
a metering device intended to detect and measure the presence of a flammable gas based on how close the gas concentration is to the lower flammable limit of the calibration gas.

FIGURE 3-1

The instrument shown in the center of this picture is a combustible gas detector combined with a multiple gas detector. Surrounding the gas detector is calibration equipment for use with the detector.

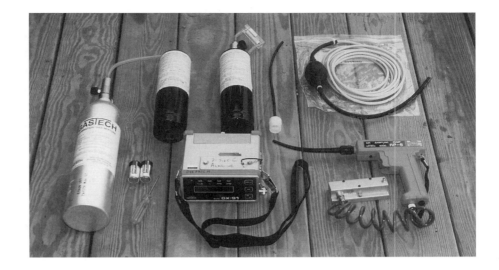

FIGURE 3-2

Combustible gas detectors indicate how close the gas concentration is to the lower flammable limit. A reading of 100% indicates that the concentration is at least at the lower flammable limit. This meter is showing a 0% LEL reading.

Figure 3-2. The CGI, also called an explosimeter, measures what percentage of the LFL is present in the atmosphere. A simple explanation of how a combustible gas indicator operates is that it is an electric circuit known as a Wheatstone bridge. One side of the Wheatstone bridge circuit is in the control chamber and the other side is in the test chamber. When an atmospheric sample is drawn into the meter, it comes in contact with a heated wire in the test chamber. As the sample is heated in the test chamber, it heats the one side of the Wheatstone bridge and the electric circuit changes resistance causing an electrical imbalance which can be read with a meter. The more out of balance the circuit, the greater the meter reading and the higher the percentage of LFL present.

SAFETY
Combustible gas indicators do not *tell you the percentage of combustible gas in air!*

There are some important questions to consider about a combustible gas indicator. First, what does the instrument measure? Combustible gas indicators do *not* tell you the percentage of combustible gas in air! They do not tell you how close the gas concentration is to the LFL. Recall from the discussion about flammable or explosive ranges in chapter 1 the lower flammable limit, the upper flammable limit, and the flammable range. Below the LFL, the gas mixture in air is too lean to burn because of insufficient fuel. Between the LFL and the UFL is the flammable range. At this point the fuel and oxygen mixture will support combustion and, if an ignition source is present, the mixture will ignite and burn. Above the UFL there is too much fuel.

Second, the combustible gas indicator gives a reading in percentage, but the question is a percentage of what? The percentage of the lower flammable limit, as shown in Figure 3-3, is the correct reading. For example, a reading of 25 percent would indicate that the meter detected a level of combustible gas that was one-fourth of the way to the lower combustible limit. A meter reading of 100 percent would indicate that the meter had reached the lower combustible limit. Importantly, a meter reading of 100 percent would not mean that the concentration of the gas in air was 100 percent. People who are not familiar with combustible gas detectors would believe that the meter reading reflects the concentration of gas in air. This assumption is wrong and could lead to a mistaken impression of the hazard.

Third, not all gases have the same flammable range or the same lower flammable limit. Be sure to know the flammable range and the lower flammable limit of the gas used to calibrate the detector. Methane, a common gas used for calibration, has a lower flammable limit of 5 percent methane in air. Pentane is also used to calibrate combustible gas detectors. Pentane has a low flammable limit of 1.5 percent pentane vapor in air, thus a meter calibrated using pentane would be more sensitive and give a greater percentage of the LFL reading in the same flammable gas/air atmosphere than other types of gas.

Fourth, in addition to knowing the gas used to calibrate the CGI, you need to know something about the gas you are trying to detect. If you are using a CGI calibrated on methane (LFL 5 percent) and the gas

FIGURE 3-3

Combustible gas detectors indicate how close the gas concentration is to the lower flammable limit of the gas with which the meter is calibrated.

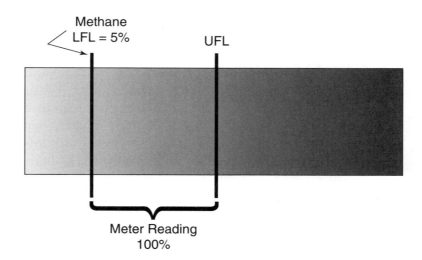

| NOTE: The CGI is meant to detect the LFL of the gas it is calibrated with.

▶ **action limit**

the highest percentage of combustible gas that is detected by a combustible gas detector which is considered as the point at which people should leave the area.

you are trying to detect, whether you can identify it or not, has an LFL of 2 percent in air, will the meter be able to effectively tell you when you have reached the 100 percent mark on the CGI? Unfortunately, no because the CGI is meant to detect the LFL of the gas it is calibrated with, as shown in Figure 3-4. If the LFL of the calibration gas is 5 percent and you are dealing with a gas whose LFL is 2 percent, you will be well into the flammable range before the CGI reads 100 percent. To compensate for this situation, an **action limit** of any CGI reading of 10 percent should be part of your emergency response procedure—any time you reach a 10 percent LFL reading on the CGI, all activities will stop in the area containing the flammable gas or vapors. You must then remove the hazard by some means (e.g., ventilation) and bring the CGI reading below 10 percent. This process serves two purposes: (1) because the calibration gas may not match the gas you are detecting, you are building in a safety factor to make up for the limitations of your equipment, and (2) if you are detecting flammable gases, you may have a dangerous situation when you reach the 10 percent limit and you must work to remove the hazard.

Finally, you must be aware that certain meters can detect a 100 percent LFL reading and then go immediately back to reading 0 percent LFL, which is more common with older meters and is caused by using the meter in an atmosphere above the UFL of the calibration gas. The meter has detected an atmosphere of 100 percent LFL and once the gas being drawn into the meter goes above the UFL, the gas is too rich to combust within the meter and heat the wire. The result is that the meter is fooled into reading 0 percent because the Wheatstone bridge circuit is now back in balance.

When using a combustible gas indicator, or any metering device, you must sample the area at different levels due to the difference in vapor density of different gases. Gases that are heavier than air can be expected to collect at lower levels, whereas gases that are lighter than air can be expected to collect at upper levels, as shown

FIGURE 3-4

For a CGI calibrated on methane, a meter reading of 50% would indicate a gas concentration equal to 2.5% methane in air.

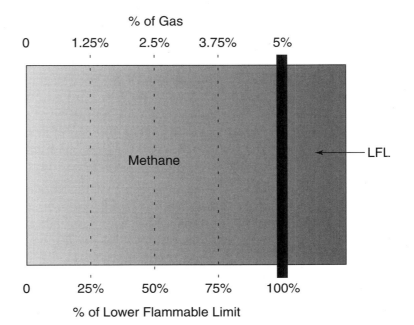

FIGURE 3-5

Depending on the vapor density of the gas you are attempting to monitor, the gas concentration may vary within the space. A gas which is heavier than air will tend to be more concentrated at the bottom of the space. The reverse would be true for a gas that was lighter than air.

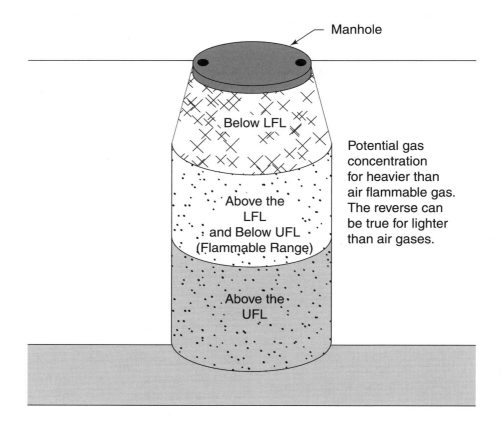

● **SAFETY**
You must meter the space at the top and bottom as well as the intermediate levels.

in Figure 3-5. If you only meter the space at the waist-high level, you can easily miss gases that are more concentrated above or below waist level. To avoid this problem, you must meter the space at the top and bottom as well as the intermediate levels.

Using a CGI and interpreting the data given can often cause confusion. To help you understand the information you are being given, consider the following scenarios:

1. A combustible gas indicator is calibrated using methane which has an LFL of 5 percent gas in air. You know that natural gas is present in the confined space and you have detected a meter reading of 50 percent. Does this tell you that you have 50 percent methane in air or that the atmosphere is 50 percent of the way to the lower flammable limit?

    The meter is telling you that you have reached a point where the concentration of the gas in air has reached 50 percent of the LFL. In fact, if you had reached a level of gas that was 50 percent gas in air, you could get a meter reading of 100 percent because it would be above the UFL of methane.

2. You are using a CGI calibrated for pentane which has an LFL of 1.5 percent in air. The meter you are using is set to alarm at 10 percent LFL. The meter alarm goes off and you are reading 15 percent LFL. Should you continue to work in this area or should you leave and try to remove the hazard?

    Depending on whether you identified the type of gas you are detecting with the CGI, you may not know the LFL of the gas. To simplify, when using the meter you should have an action

FIGURE 3-6

Oxygen levels that are above or below the "normal" range of 19.5% to 23.5% provide problems that must be addressed.

Effects of O$_2$

- ❖ **Above 23.5%—materials can ignite easily and will burn rapidly**
- ❖ **23.5%—OSHA definition as oxygen enriched**
- ❖ **21%—normal**
- ❖ **19.5%—OSHA definition as oxygen deficient**
- ❖ **17%—some muscular impairment, increased respiratory rate**
- ❖ **12%—dizziness, headache, rapid fatigue**
- ❖ **9%—unconsciousness**
- ❖ **6%—death within a few minutes**

limit of 10 percent, which means that any time you get a 10 percent reading on the meter, you will suspect that the atmosphere is hazardous and take steps to remove the hazard. Even with a meter calibrated on a gas such as pentane and with its low LFL, you must accept that a 10 percent meter reading is a strategic factor which will limit or affect your actions.

Combustible gas indicators are valuable tools in helping you to determine the safety of the atmosphere you are metering. Because the CGI actually combusts the sample being drawn through the meter, it is important that the oxygen content of the atmosphere is between 19.5 percent and 23.5 percent. At oxygen levels above or below that range, the CGI will not be able to give an accurate reading due to the increased or decreased combustion that will occur.

OXYGEN MONITORING EQUIPMENT

Air normally contains 21 percent oxygen. Increasing or lowering the oxygen content of air can have serious effects on both human life and the fire hazards presented by certain fuels. The oxygen content of air within a confined space can be affected by several means. A confined space could have been closed for a long time and something as simple as decomposition of organic materials such as leaves, grass, or septic waste by bacterial action could lower the oxygen content of air within the space. Other times the oxygen content within the space might have been lowered by something that was introduced there such as nitrogen gas for inerting, welding within the space, or painting or coating the inside of a space. Regardless of how the oxygen content was lowered, any time the oxygen content of a space reaches 19.5 percent or less, the atmosphere is to be considered oxygen deficient, as shown in Figure 3-6. No one should be allowed to enter an oxygen-deficient space without wearing a positive pressure self-contained breathing apparatus or a positive pressure–supplied air respirator.

Oxygen-enriched atmospheres—those above 23.5 percent oxygen in air—are also dangerous. Although an increase in the oxygen level will not lead to a person being overcome, it can lead to a situation in which materials normally difficult to ignite or burn can become com-

SAFETY

Regardless of how the oxygen content was lowered, any time the oxygen content of a space reaches 19.5 percent or less, the atmosphere is to be considered oxygen deficient.

bustible and burn quite rapidly, including normally safe items such as electrical equipment, high flash point liquids, and clothing.

To detect the level of oxygen within an area, use an oxygen meter. This meter functions by the atmospheric air reacting with an electrolytic material contained within the meter. As the chemical reaction occurs, an electric current is produced and the current is read on either a digital readout or on a gauge. Oxygen meters are fairly straightforward, unlike the combustible gas indicator, in that the reading is the percentage of oxygen by volume in air. Therefore, a 20 percent meter reading indicates that the air within the space being monitored contains 20 percent oxygen. To correctly monitor the atmosphere within the space, you must monitor at different levels to compensate for the differences in vapor density of different gases.

SPECIFIC GAS MONITORING

In addition to monitoring for combustible gases and oxygen content, it is possible to monitor for specific gases, as shown in Figure 3-7. It is even possible to purchase detection devices that simultaneously monitor for combustible gases, oxygen content, and several other specific gases, as shown in Figure 3-8. Specific gas monitoring instruments are generally designed to give readings in **parts per million (ppm)**. One ppm is equal to 1/10,000 of 1 percent, and is an important difference from the meter reading of CGI and oxygen meters which measure readings in percentages. The reason why these gas-specific meters measure in parts per million is that the gases they are designed to detect and measure are toxic in very low concentrations. Such gases as carbon monoxide and hydrogen sulfide are considered **immediately dangerous to life and health (IDLH)** at 1,500 ppm and 300 ppm, respectively. However, both gases are also flammable—carbon monoxide has an LEL of 12.5 percent and hydrogen sulfide has an LEL of 4.3 percent. Because the gases are dangerous to life at such low levels, but do not pose a fire hazard until they are at 125,000 ppm (12.5 × 10,000) and 43,000 ppm (4.3 × 10,000), a combustible gas indicator would tell you that there was a serious fire hazard, but it would probably be too late since anyone in the contaminated atmosphere would be dead from the toxic effects of the gas. This example may seem like mixing apples with oranges, but it serves to show that a gas may be both toxic and flammable and that attempting to use a meter designed for one purpose, such as a CGI, will not work overall.

Another type of monitoring device is a colorimetric tube. These glass tubes, commonly called Dräger tubes or Sensidyne tubes, use a chemical agent in the tube which reacts with specific gases or vapors, as shown in Figure 3-9. When the gas is drawn into the tube by a pump, the chemical agent reacts and turns color. By measuring the length of the color change against a scale, usually on the tube, you can determine the concentration of gas or vapor in the air. There are limitations to the use of these types of tubes. First, you must know or suspect the presence of a specific product so that you can select the correct tube. Then you must use the tube in a pump supplied by the manufacturer and make sure that you draw the correct amount

▶ **parts per million (ppm)**

the number of units of a particular material occurring in a total volume of one million units. One part per million is the equivalent of 1/10,000 of 1 percent.

▶ **immediately dangerous to life and health (IDLH)**

the maximum level to which one could be exposed and still escape without experiencing any effects that may impair escape or cause irreversible health effects.

FIGURE **3-7**

This is a meter which is designed to monitor carbon monoxide levels. It is not intended to detect and measure any other gas.

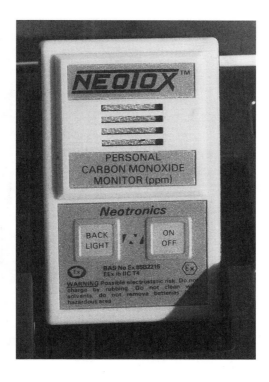

FIGURE **3-8**

The meter shown here is designed to measure combustible gases, oxygen and hydrogen sulfide.

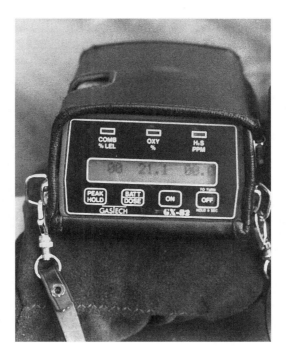

of air through the pump. To draw through the proper amount, the pump must be in good working condition, without leaks, and you must pump the device using full strokes from a manual pump. Each compression and refilling of the pump takes time and you must count the strokes. Once completed, the accuracy of your reading is ± 25 percent, meaning that if you get a reading of 10 ppm you could actually be at 7.5 ppm or 12.5 ppm. If the IDLH is 11 ppm, are you over or below its limit?

FIGURE 3-9

Colorimetric tubes are designed to be used to detect specific chemicals. These tubes are used by drawing a sample of air through the tube and then noting any color change.

▶ **corrosive**

a material that can be acidic or basic and, because of those properties, can damage human skin or rapidly corrode metal.

▶ **logarithmic scales**

scales, such as the pH scale, in which a change between whole numbers represents an exponent of the power to which the change will be raised.

▶ **pH paper**

a basic detection device consisting of pH-sensitive paper that will change colors when exposed to an acid or base. Some pH papers are designed to change color in proportion to the pH level of the material to which they are exposed.

▶ **pH pen**

a monitoring device, resembling a pen, which gives a direct reading of the pH of the material to which it is exposed.

Specific gas monitoring devices are not limited to direct reading instruments. Devices such as home carbon monoxide detectors give a warning when the carbon monoxide level reaches a certain point. The problem is that you do not know how much past the alarm point the carbon monoxide levels are. Some detection devices on the market actually monitor many different gases and sound an alarm when a gas is detected. The problem is that these devices do not give direct readings, but rather sound an alarm. Devices that are not direct reading should not be used for confined-space rescue work.

pH DEVICES

Materials that are **corrosive** can be classified as either an acid or a base (also called alkali or caustic). To measure the strength of an acid or base, you use a logarithmic pH scale ranging from 0 to 14. On this pH scale, 7 is neutral and materials having a pH of less than 7 are acidic and materials above 7 are basic, as shown in Figure 3-10. To understand the scale you should know that **logarithmic scales** are based on a multiplier of ten. Every time the pH scale changes by one whole number, the strength of the acid or base increases or decreases by ten times. A simple example of this is an acid with a pH of 1.0 that would be ten times as strong as an acid with a pH of 2.0.

Corrosive materials that have a very low or very high pH are strong acids and bases. These corrosives on contact are capable of damaging human tissue, metals, or other materials. To qualify the degree of hazard that emergency responders might face from a solid, liquid, or gaseous corrosive, you need to determine the pH of the material by using either a pH meter or other pH detection equipment such as **pH paper.** Regardless of the method used to determine the pH of the material, your detection equipment will have to come in direct contact with the corrosive material. Therefore, you will need to provide protection from the material for the person using the detection equipment.

Meters that detect pH are simple, direct reading instruments, such as a **pH pen** which looks like a ballpoint pen, or a more sophisticated

SAFETY

After you have used these meters or pH papers you should be aware that some of the corrosive material can remain on the meter probe or pH paper.

SAFETY

In fact, you should monitor the atmosphere outside of the confined space as you approach to be sure that contaminants are not venting from the area and creating a hazardous atmosphere where rescuers will not expect it.

FIGURE 3-10

A pH chart showing the pH range of some common materials. The range is from 0 (very acidic) to 14 (very basic).

pH OF SOME COMMON SUBSTANCES

	14	
lye		
	13	
household ammonia	12	
		BASIC
	11	
lime water		
	10	
borax		
	9	
baking soda		
	8	
blood		
milk	7	NEUTRAL
rain	6	
black coffee	5	
tomatoes	4	
soda		
	3	ACIDIC
lemon juice	2	
gastric fluid	1	
	0	

pH meter, as shown in Figure 3-11. The probe is placed into the material and a reading from 0 to 14 is shown on the readout of the instrument. pH paper strips, shown in Figure 3-12, are placed in contact with the corrosive material and they then react to change color. Some pH papers will simply turn red or blue to indicate the presence of an acid or base, but others will turn varying shades of colors which are then matched with a chart to show the pH. After you have used these meters or pH papers you should be aware that some of the corrosive material can remain on the meter probe or pH paper.

It will be necessary to avoid handling these items if they are contaminated with a corrosive material.

UNDERSTANDING MONITORING EQUIPMENT READINGS

At any confined-space incident you need to monitor the atmosphere of the actual confined space. In fact, you should monitor the atmosphere outside of the confined space as you approach to be sure that contaminants are not venting from the area and creating a hazardous atmosphere where rescuers will not expect it. You must always monitor for combustible gases and oxygen content, and also preplan with local facilities that have confined spaces to help determine what other gases to expect. A proper evaluation or size-up of the emergency can also help. Is there a confined-space permit present? Does it identify what work was being done in the space or what products may have been in the

FIGURE 3-11

This device is a pH meter, and it is designed to be used in a laboratory. You could bring a sample of the product to the lab, but in most cases this would not be practical. That is one of the values of a portable meter.

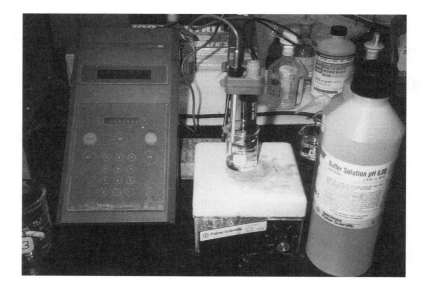

FIGURE 3-12

pH paper is practical to use in the field, but you must know the type of paper you are using and what a color change indicates.

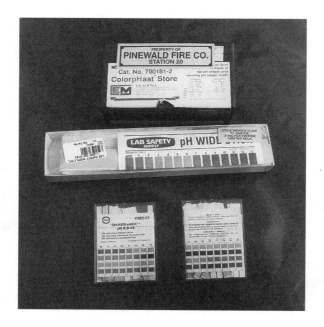

space? What about other workers—were they briefed about any hazards that might be present? Are there markings on the vessel or tank (the confined space) as to what it contained before entry? Did anyone report a characteristic odor or taste (you should not confirm this by smelling or tasting) from the product or contaminants in the space? Are there any other indications as to the atmospheric hazards? You can expect certain gases in below-ground spaces such as sewers and septic tanks and above-ground spaces such as silos. Combustible gas indicators tell you how close you are to the LFL of the gas on which the meter is calibrated. If you reach the action limit, have a plan as to what you can do to control the hazard. In addition, all instruments must be maintained, including periodic calibration. When was the last time the CGI was calibrated?

Oxygen meters will tell you the oxygen content of the atmosphere you are monitoring. Do you know the limits for oxygen-deficient and oxygen-enriched atmospheres? What about the hazards each type of oxygen atmosphere presents regarding the danger of fire or the need for respiratory protection?

Specific gas monitoring instruments can read the levels of the gas for which they are intended, but you must also match the gas to the meter. In addition to specific gases, are you aware of any interferants that might affect the meter? Interferants basically trick a meter into believing that the specific gas is present, because they react with the meter much the same way that the specific gas would. You may have a carbon monoxide meter in alarm because it detected the vapors coming off a product such as benzene in gasoline. Know what materials can cause a false reading on your gas meter before you attempt to use it at an emergency.

Colorimetric tubes have real value in that they can confirm the presence of a specific material. Beyond that they have some serious limitations. Like all monitoring instruments, most are used by emergency responders and have been adapted from other fields such as safety and industrial hygiene.

► **radio frequency (RF) interference**

electromagnetic interference caused by a signal generated by an electrical device.

Other items that should be addressed include **radio frequency (RF) interference,** setup of the instrument in clean fresh air, manufacturers' instructions, monitoring at all levels, calibration, and training. The use of radios in close proximity to some electronic monitoring devices can cause a false alarm. Do not transmit on a radio that is close to a monitoring instrument. Keep them separated to avoid RF interference and check with the manufacturer to see if this is a known problem. Any time you are going to use a monitoring instrument, it should be set up (turned on, warmed up, and zeroed if needed) in fresh air to avoid a contaminated background reading as the acceptable starting point. Read and follow all manufacturers' instructions. They know the ability and limits of their equipment. They are also liable for the equipment's performance and want to protect you from doing something foolish with it. Because not all atmospheric hazards have the same vapor density, it is critical that you monitor all levels of the confined space. Simply checking only the top or bottom of an area can lead you to a false reading. Materials that are lighter than air will accumulate at the top of a confined space while materials that are heavier than air will accumulate at the bottom. If you only monitor one level you can easily miss a high concentration of the atmospheric hazard, so monitor all levels of the space—top, bottom, and intermediate levels.

The manufacturer can be a valuable source of information for training with and using the monitoring equipment. Calibration is critical to the proper performance of any monitoring equipment. At the very least, an uncalibrated monitoring instrument can give you false high readings; at the worst it can give you false low readings or no reading at all. You then assume that all persons on the rescue team are safe when in fact they are in serious danger. Failing to train with the monitoring instruments is almost as bad as not calibrating the equipment. When you train to use the equipment properly, you are calibrating the operator. Know the limitations of your equipment. Know what it is telling you when you get a reading. Know how to use the information you develop by monitoring the confined space to protect rescuers and victims.

■ SUMMARY

Monitoring instruments will allow you to characterize the atmospheric conditions at a confined-space incident; however, you must know what your instruments are capable of detecting, how to interpret the information the instrument is giving you, and what the information will mean to your rescue operation. You must also be aware of the limitations of the equipment in that the accuracy of the readings, the need for routine maintenance such as calibration, and the specific instructions from the manufacturer will all affect the performance of the equipment. If you are to trust the readings you get from the equipment, you must know how to properly use and maintain it. The time to learn to use your monitoring equipment is long before you have a confined-space incident. Through training and practice with the equipment, you will develop proficiency. Remember to set up your equipment in a clear area so as to avoid getting a false reading from contaminated air, to monitor at all levels of the confined space, and to continuously monitor as long as an atmospheric hazard has the potential to affect your operations.

■ REVIEW QUESTIONS

1. You are using a combustible gas indicator and have a meter reading of 8 percent. Does the reading mean that you have 8 percent combustible gas in air or that the atmosphere is at 8 percent of the lower flammable limit?

2. While using an oxygen meter, you get a reading of 28 percent. Is this meter reading oxygen deficient, oxygen enriched, or within the normal range? What hazards would you face from this atmosphere?

3. A multiple gas detector (combustible gas, carbon monoxide, hydrogen sulfide, and oxygen) is giving you a reading of 15 for carbon monoxide. What following unit of measure is the meter reading: percent of gas in air, percentage of the LFL, or parts per million?

4. A pH meter reads 1 on the pH scale. Would this reading indicate an acid or base? Would the reading indicate a strong concentration or a weak concentration of the acid or base?

5. Combustible gas indicators tell you the percentage of combustible gas in air. True or false?

6. The CGI is meant to detect the ____ of gas with which it is calibrated.

7. What levels of a confined space should you monitor before entry?
   a. Top
   b. Bottom
   c. Middle
   d. All of the above

8. What does IDLH mean?

CHAPTER

4 Lockout/Tagout

OBJECTIVES

After completing this chapter, the reader should be able to identify the purpose of locking out and tagging out energy sources and equipment, including the following energy sources:

- electrical
- hydraulic
- pneumatic
- chemical
- thermal
- gravity

The reader should also be able to:

- define how preplanning can assist in identifying the need for lockout/tagout
- describe how a hazard and risk assessment applies to lockout/tagout requirements
- identify examples of equipment and methods that can be used for lockout/tagout

INTRODUCTION

How many times have you seen the cartoon with the roadrunner and coyote? Invariably the coyote creates some device to capture the roadrunner. The contraption may involve a spring, a catapult, or a rope that is holding a heavy object. The roadrunner stops, takes the bait, but escapes. When the coyote goes to investigate why his trap failed to work, the device operates and the coyote becomes the victim of his own invention. Sounds funny when you think of it as a cartoon, but what about when the situation involves a confined-space accident where some electrical or mechanical device shares the space with the victim? Imagine that a rescuer enters the space and the energy stored in the device releases and the rescuer now becomes the victim. This situation seems grim, but it does not have to be that way. You can be smarter than the "coyote" by isolating the energy source before entering the space, which is the purpose of locking out and tagging out equipment, products, and processes.

To understand why you need to lockout/tagout certain energy sources, you need to begin by understanding the hazards that those sources present. The OSHA Standard 1910.147 Control of Hazardous Energy (Lockout/Tagout) defines an energy source as "[a]ny source of electrical, mechanical, hydraulic, pneumatic, chemical, thermal, or other energy." This definition is broad, but by developing some basic examples you will begin to understand what is meant by an energy source. You would expect to find electrical equipment in an underground electrical vault and you should be fairly familiar with the electrocution hazards that live electrical equipment poses, but what about a silo filled with corn kernels? Gravity is the energy source within the silo. If the corn starts to flow within the silo, it can engulf both rescuer and victim.

Imagine a 30,000 gallon vertical tank which has a liquid product line going into it and the line valve has not been isolated or locked out from the tank. As your rescue team works to assist a victim in the tank, a plant operator remotely opens the valve and hundreds of gallons of corrosive liquid begin to flood the tank. How would you feel about committing rescuers to a process vat in a chemical plant which had a large paddle, called an agitator, running through the middle of the vessel if you could not be sure that the agitator could not be started by accident? These examples are different types of energy sources and all of them introduce a hazard into the confined space. Controlling these energy sources is what lockout/tagout is about so that "unexpected energization or startup of the machines or equipment, or release of stored energy" will not cause injury to people, as shown in Figure 4–1.

LOCKOUT/TAGOUT REQUIREMENTS

There are several ways in which you, the rescuer, can become aware of the need to lockout/tagout energy sources. The easiest way is by preplanning expected rescue situations. If you are part of a confined-space rescue team who works for a private company, you

SAFETY
"Unexpected energization or startup of the machines or equipment, or release of stored energy" can cause injury to people.

FIGURE 4-1

A lockout control center showing some of the equipment required for controlling energy sources.

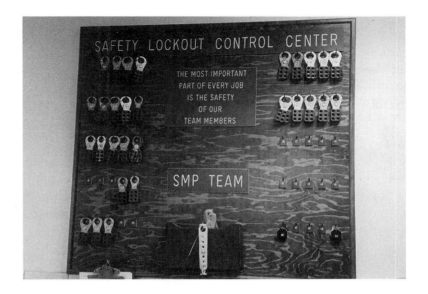

▶ **hazard and risk assessment**

determining what has happened to create the emergency, what conditions are still present or will evolve at the emergency, and then predicting what can be done to resolve the emergency.

| NOTE: You will need to seek out the confined-space entry permit, the entry supervisor, the attendant, and any other person who witnessed or has knowledge of the incident.

can easily survey the entry sites prior to issuing a permit. Identifying the need for implementing lockout/tagout procedures is a requirement of the confined-space entry permit. If you are a member of a public sector response team, you can still do a certain amount of preplanning, especially if your team is the designated rescue team for the confined-space site. Preplanning allows for an indepth, accurate assessment of the situation before an emergency. Preplanning eliminates the sense of urgency to make decisions and commit rescuers. Even if you preplan a confined space, you still need to ensure that conditions have not changed in the space since your last inspection. But what if this confined-space accident happens in an unfamiliar area and you have had no chance to preplan? Certainly a **hazard and risk assessment** is needed, but where do you start? How do you gauge the accuracy of the information you are receiving? How do you identify potential energy sources which need to be locked out, as shown in Figure 4–2. How do you identify and control sources of residual or stored energy such as a compressed spring or a charged electrical capacitor?

This chapter is devoted to lockout/tagout and any discussion of hazard and risk assessment will be confined to determining the need to lock out hazardous energy sources. Any hazard and risk assessment must begin with three simple questions: What has happened? What is happening now? What can you expect to happen in the near future?

The answer to the first question is EMS personnel look for the mechanism of injury and firefighters try to locate the seat of a fire. You are going to do the same thing except that you are looking at the victim, the confined space, and the area surrounding it. You will need to seek out the confined-space entry permit, the entry supervisor, the attendant, and any other person who witnessed or has knowledge of the incident. Then you will have to figure out how all of these pieces of the puzzle fit together. Is an energy source or mechanical equipment involved in the accident? If so, then how does that factor affect your rescue operation? You also need to keep in mind that where me-

FIGURE 4-2

Product stored in this hopper is intended to be released through the chute at the bottom. The source of the energy moving this product is gravity and it must be thought of as stored energy.

chanical devices are present and are entrapping the victim, you may not have the skills necessary to rescue the victim. At this point you may have to call for additional help from a team that is specifically trained in removing victims from machinery.

The second question is important for your size-up. If an energy source was involved, was all of the energy released or does the potential exist for more problems to occur in the confined space because of any remaining energy? If your victim was injured or trapped by an energy source, can you control that source from outside the space? Is the energy source affecting the victim or the potential for rescue? A difficult question to ask and to answer is if this emergency is a rescue or a victim recovery. If the energy source makes it obvious that your victim is dead, your whole operation changes from a rescue to a recovery and additional steps may need be taken to protect emergency responders.

The third question is a little like fortune-telling, but based on your experience and knowledge you can make fairly sound predictions. A victim trapped in a confined space with water rushing into it would be expected to drown when the water reached head level. Likewise, a victim caught under a heavyweight object would be expected to have crushing injuries. That kind of prediction does not require a crystal ball. At times, however, your decision and course of action may at best be only a guess as to the outcome: You expect one thing to happen and suddenly something completely different happens. Remember, this chapter is devoted to lockout/tagout, so the discussion here will be limited to such. When making predictions about what will happen, you need to be aware of some of the methods of controlling energy sources. Lockout/tagout means that you physically lock a valve, switch, or other energy-controlling device.

▶ **manually operated electrical circuit breakers**

circuit breakers within an electrical circuit that are designed and intended to be safely operated by direct manual manipulation.

▶ **disconnect switches**

an electrical switch designed to isolate the electrical source from the equipment that it powers by disconnecting the power supply from the equipment.

▶ **line valves**

valves in the piping which allow product to travel into a tank, vessel, vat, or other confined space.

▶ **latches**

fastening devices that consist of a bar that falls into a notch to prevent opening or operation of the object they secure.

▶ **chains**

flexible series of joined links or rings, typically of some type of metal.

▶ **chocks**

blocks or wedges designed to prevent motion of an object they are placed into or under.

▶ **blank flanges**

flanges that have no opening in them and are meant to block the flow of a product past the flanges.

▶ **blocks**

energy-isolating devices meant to stop or obstruct the flow of hazardous energy or products.

▶ **bolted slip blinds**

blinding devices that are meant to be bolted directly to a flange to stop or obstruct the flow of a product.

After the device is locked, it is tagged so that others will know that someone has isolated that device and will not attempt to operate it. No one attempts to remove the lock and tag except for the person who placed the lock. Examples of devices that can be locked include **manually operated electrical circuit breakers, disconnect switches,** and **line valves** using specially designed lockout/tagout equipment, as shown in Figure 4–3. You may also use **latches, chains,** and **chocks** to secure energy sources. In addition to using a lock, it is also possible to use **blank flanges, blocks,** and **bolted slip blinds,** as shown in Figure 4–4. Whatever device you use for lockout/tagout, keep in mind what you are trying to accomplish: You want to secure a potentially hazardous energy source to protect both the rescuers and victims and you want to be sure that it is effective and cannot be easily defeated, as shown in Figure 4–5. If you were to consider stationing a person at the switch, valve, or other controlling device instead of using another method of lockout/tagout, you would be taking a large risk. Mechanical devices do not leave their position or open the device by accident because of miscommunication. Using a person for your lockout/tagout device should be a method of last resort and done under only the most extreme conditions.

As part of your confined-space response equipment you should carry some basic lockout/tagout equipment. The need for more sophisticated equipment can be determined through preplanning, and site-specific equipment may be on-site. At the very least, it is possible to disconnect and misalign pipelines, belts, chain drives, and mechanical linkages.

Factors critical enough to have caused, modified, or affected the outcome of an emergency are strategic factors. Strategic factors demand that you either change the factors or, if they cannot be changed, work within the limits set by these factors. You are still answering the same three questions: What has happened? What is happening now? and What is expected to happen? When the presence of energy sources that must be locked out become a strategic factor, you must determine how those factors affect your operation. If you had the opportunity to preplan the confined space, you should have anticipated these factors and planned for them. If you did not have the chance to preplan, you need to look at the problem, your equipment, and your standard operating procedures (SOPs) to determine if you have the capability to handle it. Should you find that your equipment or your SOPs will not work, you have to develop a plan that will work. Keep in mind that you will not be able to handle every emergency on your own. Preplanning does not have to be limited to the confined space— locating and knowing how to secure additional resources is also an effective means of preplanning.

As an emergency responder, all concerns about lockout/tagout can seem confusing. It is important that you look at the whole picture and not allow tunnel vision to occur. Prior to attempting a rescue in a confined space or any space around or near machinery and equipment, you must determine if there is a need for lockout/tagout. Ideally, you were able to preplan the accident site, but in the real world you may have little information regarding the area and the

FIGURE 4-3

Lockout/tagout equipment is intended to control various types of equipment. Shown here are some locking devices for valves, switches, and electrical equipment. The confined-space rescue team should have some basic lockout/tagout equipment as part of their equipment.

FIGURE 4-4

The device shown in this photo is called a spectacle blind since it resembles a pair of spectacles. One side of the device is solid (blind) and intended to stop the flow of product through the pipe while the other side is open.

 NOTE: As part of your confined-space response equipment you should carry some basic lockout/tagout equipment.

NOTE: It is possible to disconnect and misalign pipelines, belts, chain drives, and mechanical linkages.

hazards it may pose from hazardous energy sources. When you are faced with little information, you must begin developing that information as quickly and accurately as possible, which means that you will have to question anyone who witnessed the accident and those working at the site. Failing to control these energy sources can kill or permanently disable your personnel. Lockout/tagout controls these hazards. Sources of energy and power must be released and controlled before attempting a rescue.

Consider two sample incidents. Begin by looking at the strategic factors and lockout/tagout considerations.

FIGURE 4-5

This pipeline is locked out by breaking the line and placing a blind flange on the open end of the pipeline.

FIGURE 4-6

Warning sign indicating the presence of automatic starting equipment. Automatic starting equipment can start without warning.

| NOTE: Look at the whole picture; do not allow tunnel vision to occur.

 **SAFETY**
When you are faced with little information, you must begin developing that information as quickly and accurately as possible, which means that you will have to question anyone who witnessed the accident and those working at the site. Failing to control these energy sources can kill or permanently disable your personnel.

At the first incident you have been called to a confined-space accident where the space is an underground generator room at a computer operations facility for a large bank. Workers were inside the room using a solvent to clean up spilled diesel fuel, when one employee complained of feeling dizzy and then collapsed. A second worker alerted the attendant and was pulled to safety using a harness and tripod. The worker who is now trapped in the confined space is wearing a harness but is not connected to a retrieval line because fuel lines and an electrical conduit divide the space. The worker disconnected his retrieval line to get around the conduit. There are no product lines or other piping connected to the area and you do not see any tags or locking devices indicating that the equipment has been locked out or tagged out. A large sign at the entrance to the space reads "**Danger—Automatic Starting Equipment**," as shown in Figure 4–6.

Viewing the incident should give you an idea about what has happened. You should also be aware of what is happening now or at least know where to start looking for answers. Predicting what is going to happen should also be a straightforward task. If the atmosphere of the space can be made safe to enter and you have the respiratory equipment and personnel available, you will probably attempt to rescue the victim. Recall the warning sign, however, and locate a knowledgeable person from the facility to tell you what the automatic starting equipment is and how it starts. You will also need to know how to lock out and tag out the equipment so that it does not start while you are attempting to rescue the victim. Because this incident is in a computer facility and the space is a generator room, a power failure may cause the equipment to start; however, this equipment is critical to the continued operations of the facility and a power failure could cost millions of dollars a minute if the computers go down. The generator might be on a timer that will automatically start the equipment to test and exercise it to ensure reliability. The question now becomes, "If the generator is the automatic starting equipment, should you have it locked out and tagged out to prevent it from starting?" The answer is yes. At the very least this equipment may be extremely noisy when it is operating. As a result, rescuers may not be able to hear or communicate and they may even have their hearing damaged. At the worst, this equipment could be a potential ignition source for flammable vapors or it may have moving parts that could endanger the rescuers and the victim. Have the equipment locked out and tagged out for the duration of the rescue, even if the facility manager is reluctant due to the potential loss should the power fail. Eliminate the potential hazard and the strategic factor goes with it.

For the second incident, imagine that you have been called to a manhole that leads to an underground storm sewer line. In the confined space is a worker trapped at the waist by a pile of bricks from the sidewalls of the sewer, as shown in Figure 4–7. The space is vented, and the victim is conscious but not fully alert and attached to a full body harness and retrieval line on a tripod. Water is entering the space from the soil at the area where the bricks fell from the wall and the water is just below waist level on the victim. The location of this manhole is 50 feet from a large river which empties into the ocean several miles away. The river at this point is still subject to the tide from the ocean. As you approached the scene the entry supervisor gave you a copy of the entry permit. The permit indicates no need for any lockout/tagout items. The supervisor tells you that this was an unexpected problem, because the manhole was pumped dry and there is a one-way valve on the end of the storm sewer pipe where it enters the river. In the words of the supervisor, "Everything was dry and secure before we went into the space. It just gave way after we'd been in there for an hour."

First consider what happened and, since this chapter deals with energy sources, if any energy sources are present. Think about what is making the water move. Gravity is causing the water to flow through the ground and seek its own level. Because you are 50 feet from the river, the level of the collapsed area in the manhole might

FIGURE 4-7

Illustration of a confined-space accident.

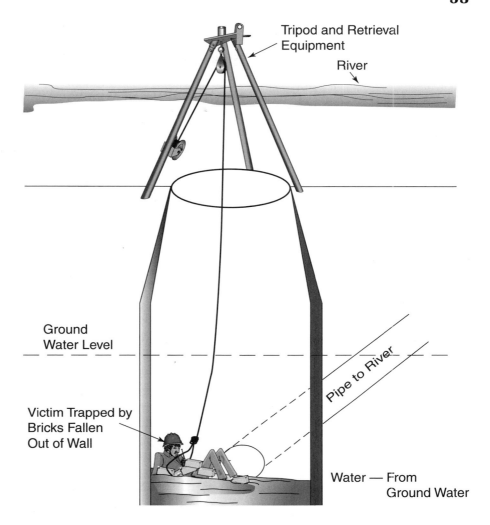

actually be below water level of the river. As the water was pumped from the manhole, water pressure built up against the walls of the manhole. If the manhole was full of water to the level of the river, the pressure inside the manhole and outside the walls would have been equal and the walls would not have collapsed. This fact helps to determine what has happened.

What is happening now as you perform your assessment? The water continues to fill the space and the danger to your victim increases with every minute. As to what will happen in the future, that depends on your plan and on other factors beyond your control. First, this is a tidal river. Is the tide coming in and increasing the level of the river or is the tide going out and decreasing the level of the water? If the tide is coming in, you can expect the flow of water to increase as the water level rises. You may even have additional collapse of the wall as the water level rises. If the tide is going out, the water flow may not worsen, but you may still be below the level of the river's surface. The water has become a strategic factor. Can you stop the flow of water? Probably not because you are trying to turn back millions if not billions of gallons of water from the river and the ocean. Can you control the flow of water? You could set up a pump that would draw the water off at a rate equal to that which is coming in—not exactly a typical lockout/tagout, but it is certainly some level of control of a source of energy which threatens the victim and the rescuers.

Regardless of whether you can control the water level, you should be able to predict what is going to happen. If you do not control the water, the victim will drown. If you can control the water level, you will be able to prevent the victim from drowning and you may be able to have at least one rescuer enter the space and free the victim. At that point, you can then remove the victim from the manhole in the most appropriate manner.

What has become important to your rescue operation in these two scenarios is that you were able to control the sources of a potential hazard—one by a simple means of lockout/tagout and another by less conventional means. In each case, control of potentially hazardous energy sources was a critical factor. For the first example you were able to eliminate the hazard and with it the limiting factor. During the second example you were only able to exercise a measure of control of the hazard. Had the pump been unable to keep up with the water entering the space or had the pump failed, the limiting factor would have become extremely critical and could have endangered both the victim and the rescuers. Remember, eliminate or control the potential hazardous energy sources. Exercise the highest level of control by physically locking out the energy source and tag the lockout devices so that you can keep track of them.

NOTE: Remember, eliminate or control the potential hazardous energy sources.

■ SUMMARY

Confined spaces can contain many different hazards not the least of which are physical hazards. Before attempting to rescue a victim you must be aware of all the hazards within the space. For hazards that present the danger of injury or death through the sudden or unexpected release of an energy source, you must control or neutralize that energy. Control of the energy sources require that you lock out and identify, by tagging, the flow of that energy. Devices which lock, block, divert, or otherwise keep the hazard from entering the confined space will allow you to control the problem, but you must be prepared in advance by having the equipment available to you. Preplanning is always an advantage and will save time in assessing the problems that you face, but not all of your incidents can be preplanned. When you are faced with a confined-space rescue problem, you must assess the situation thoroughly and ensure the safety of the rescuers and the victims. When risks are present, you must seek to reduce or eliminate those risks. Lockout/tagout is one method of controlling the risks.

■ REVIEW QUESTIONS

1. Identify how potential energy sources can affect operations within a confined space. For each source you identify, explain how it can endanger rescuers.

2. If you were given a facility containing confined spaces to preplan, what items would you want to identify for lockout/tagout? Would you also consider asking questions about the facilities lockout/tagout program and equipment?

3. Hazard and risk assessment requires you to identify what has happened, what is happening now, and what will happen in the future. If you determine what has happened to cause the emergency and you can see what is happening as you size up the emergency, how will this allow you to develop an action plan to control energy sources?

4. Identify four basic items or methods for accomplishing lockout/tagout? Explain why one is better than the other.

5. As an emergency responder, and prior to attempting a rescue, you must look at the whole picture and not just the _____ _____.

5 Using the Incident Command System

OBJECTIVES

After completing this chapter, the reader should be able to:

- identify the purpose of using an incident command system to control and coordinate activities at a confined-space emergency
- outline the incident priorities and explain the order of implementation of the following priorities: life safety, incident stabilization, and property conservation

The reader should also be able to define the functions of the following terms and positions within the incident command system:

- command
- operations
- planning
- logistics
- finance/administration
- safety
- liaison
- public information officer
- span of control
- unity of command
- action plan
- single command
- unified command
- life safety
- incident stabilization
- property consevation

INTRODUCTION

Imagine a group of professional musicians who have formed a band. All the musicians are talented and well known. Each has a favorite song and wants to perform that song. What do you expect will happen? Will the musicians come to an agreement and perform together, or will they argue among themselves and not perform at all? Now imagine that you have been called to the scene of a confined-space accident which requires you to work with other emergency responders to rescue a victim. You are a professional, and you have joined with other professionals to perform. The question is, "Who decides how you perform together?" The answer not only affects the victim, but also can affect your safety and ability to get the job done.

There are many reasons for using an **incident command system (ICS)** to effectively manage an emergency operation. Among these are the need to address the issues of safety, unity of command, and an effective span of control. The people in charge of an emergency scene need to know *who is where, doing what,* and *when.* There must be a plan from which to operate and people must either work within the plan or advise the person in charge why the plan will have to be changed. There can only be one plan in use at a time and everyone must be aware of it and what is expected of them. How many times have you been at an emergency when there seemed to be no plan, or worse yet, two or more plans? What was the outcome? Were the needs of both the victims and the emergency responders addressed properly, or were efforts and resources wasted? Thus, some type of **incident management system** or incident command system must be in effect. You must use your talent and skills in sync with all the emergency responders.

Any system for managing an incident must begin with the first trained people who arrive at the scene of an emergency. Whether you realize it or not, when you are the first trained person at the scene, you must establish command and initiate a plan of action. As an emergency responder, you have probably received countless hours of training in your area of specialization. You feel comfortable extinguishing a fire, assessing a patient, or performing almost any other task you are trained to perform. But what happens when you are the first trained responder to arrive? Do you realize that you are in charge of the emergency? Hopefully your first reaction is to begin assessing the emergency scene and determining if you can handle the immediate situation. If you can manage the emergency with the resources you have, then you may decide to return any other emergency responders before they get to the scene. If the emergency is beyond your ability and control, call for additional resources and begin planning how you will use them. Either way, you have taken control of the emergency and established command. For a large-scale emergency there is so much to be done that you will not be performing the actual tasks but rather determining the tasks that are needed and then directing other people to perform them, as shown in Figure 5-1. Regardless of your rank in the emergency organization, you may end up being in charge of the accident. You may only need

▶ incident command system (ICS) incident management system

a recognized system for providing management of personnel, resources, and activities during emergency operations.

NOTE: For a large-scale emergency there is so much to be done that you will not be performing the actual tasks but rather determining the tasks that are needed and then directing other people to perform them.

to call for more people and equipment to assist the victims, and you may only be in charge for a short time, but you were the most highly trained person at the scene at some point and thus you were in charge. If you or anyone else failed to take charge of the emergency, what would have happened?

As discussed, the reasons for using an incident command system include safety, unity of command, and an effective span of control. The idea of safety should seem pretty simple, but perhaps not so obvious. You will be concerned about the safety of the victim(s), emergency responders, spectators, and others who might be affected by the emergency. The issue of safety is easy to address—you simply correct or control unsafe conditions and act when you become aware of them. The key to safety is prevention, but the key to prevention is recognizing unsafe conditions and/or unsafe acts. Confined spaces come with all types of unsafe conditions both inside and outside. The idea that someone must take control of the incident also implies that the same person must plan to control or correct unsafe acts and conditions.

UNITY OF COMMAND

Unity of command is a straightforward concept. There should only be *one boss* to whom anyone is held accountable. Although this seems obvious, think of the opening example of the musicians. The musicians have competing interest in getting their own song played. At an emergency scene there are competing interests between police, fire, and EMS, and between different units within each agency, as shown in Figure 5-2. Not only does someone have to be in charge but they also cannot allow people who were given assignments to be redirected by other emergency responders who believe that their need is more critical. Having more than one person giving orders at an emergency is not only demoralizing, but also unsafe because it leads to confusing or multiple plans. The end result is that the goals

> NOTE: You will be concerned about the safety of the victim(s), emergency responders, spectators, and others who might be affected by the emergency.

FIGURE 5-1

Shown here is a group of emergency responders with one person in charge and briefing the others in what needs to be done.

and objectives, which the incident commander expected to be met, may or may not be accomplished. If part of the plan places you in danger and that risk can be controlled by having people continuously monitor or vent the confined space, but those people are taken off that ventilation or monitoring by another emergency responder, then what happens to you?

SPAN OF CONTROL

Span of control is the concept that any one individual can effectively manage only a certain amount of people. The military routinely uses a practical span of control of one supervisor for between three and seven workers. In the emergency services, the ratio of one supervisor to three to seven workers is effective, as shown in Figure 5-3. The span of control is based on the supervisor's ability and on factors including the degree of hazard of the work to be performed, the ability to communicate with personnel, and the number of workers needed. During a hazardous materials emergency the entry team consists of typically two people, which is a manageable span of control. During a confined-space incident you may have to work with a span of control of one to one or one to two depending on conditions. If the confined space were tight but easily accessible, then you may only be able to fit one emergency responder into the space with the victim. Regardless of how many rescuers you use, you must maintain a manageable span of control at all times.

COMMON TERMINOLOGY

The ICS is considered an **all risk system,** meaning that it can be used at all types of emergencies. For any ICS to be effective, it must have some basic elements. All emergency response agencies must be able to understand each other's roles, terminology, and procedures.

▶ **all risk system**

an Incident Command or Management System that can be used at many different types of emergencies, including but not limited to fires, police actions, emergency medical calls, and other emergencies that threaten public safety.

FIGURE 5-2

Unity of command is the concept that each person or group of people should have only one person directing them.

Unity of Command

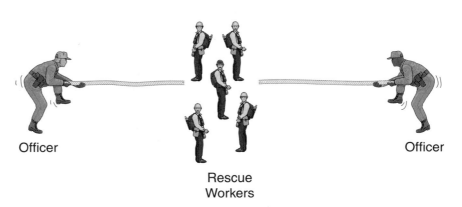

Officer

Rescue
Workers

Officer

There should only be one person
directing any group of workers.

FIGURE 5-3

The span of control refers to how many people or groups of people one person can effectively supervise and direct. Ideally, the span of control is 1 supervisor to 5 workers, but depending upon the circumstances it can range from 2 workers to 7 workers per supervisor.

1 Supervisor
or
Officer

5 Subordinates

Ideal ratio for effective span
of control is 1:5, it can
range up to 1:7.

| NOTE: Speak plain language and avoid using codes unless you are positive everyone understands what you are saying.

The police officer on the scene should know who is in charge and to whom they should address fire or EMS concerns. Simply having a police officer tell a firefighter on the scene that there is an uncontrolled pipe carrying toluene into the confined space does not ensure that the ranking fire officer will be told about it. Use terminology that a responder from another agency can understand. How many times have you used jargon from your emergency service and had someone not understand what you said? What is a wagon, or a bus? Your interpretation may not be the same as another person's, as shown in Figure 5-4. Speak plain language and avoid using codes unless you are positive everyone understands what you are saying. Procedures are more difficult to describe in common terms among different emergency agencies and they may need to be explained during an emergency. The important point here is to ensure that others know what procedures you will be using before they assist. Something as simple as rigging a hauling system can vary between agencies due to the type and amount of training they have had. Taking a few minutes to establish procedures will save having to redo it later.

The day-to-day operations of an emergency medical service, fire, or police department involve many different administrative tasks which must be accomplished for the organization to function. The people who direct those tasks are considered managers. Business managers function by planning, organizing, leading, and evaluating. Often these business managers have weeks or months to plan and organize their resources which they will then use to perform their assigned tasks. Even the time allotted for performing those tasks may be over an extended period and the evaluation may be

FIGURE 5-4

The same piece of equipment can have different names depending upon local preferences. Using a common terminology allows your message to be understood.

Nozzle
Pipe
Tip
?

Pumper
Engine
Wagon
Truck
?

quite simple. Managers may even take a vacation without needing a specific replacement in their absence.

The person in charge of an emergency scene is also a manager. That person must plan, organize, lead, and evaluate the site. However, several critical differences exist between managing a business and managing an emergency. First, the speed differs at which decisions must be made and implemented at an emergency. Then the degree of danger which threatens the victims, emergency responders, and spectators is higher. Planning at an emergency is often done with limited and perhaps inaccurate information. Often planning relies on past emergencies of similar types, experienced estimates of what was reported to you or what you observed, and then a prediction of what will happen. Organizing your tasks, resources, and needs at an emergency can be a dynamic function. You anticipate a need for more ambulances or other specialized equipment and you call for the equipment, but, it may or may not arrive as you expected. Worse yet, you do not anticipate a need and find yourself playing catch-up with the emergency and never seem to get ahead. Then your leadership function is not what you expect. Perhaps you explained to your people what you wanted them to do but they did not comply. All the while the emergency progresses nicely in spite of your best efforts.

SINGLE COMMAND AND UNIFIED COMMAND

When using an incident command system, different types of command may be needed at an incident. The simplest and most common form of command is the **single command.** Single command places one person in charge of the emergency. That person develops the action plan and tells his subordinates what he wants done. An emergency such as a fire or some basic type of police action, when only one response agency is actively involved in handling the incident, is an example of single command.

▶ **single command**

a form of command within the incident command system when a single individual is responsible for the tasks assigned to the incident commander.

► **unified command**

a form of command within the incident command system when more than one individual is responsible for the tasks assigned to the incident commander.

NOTE: The type of command that is used will depend on the incident commander's decision to use single command or unified command.

► **action plan**

a plan developed by the incident commander that establishes goals and objectives for the emergency, identifies the resources to be used, and provides the means to accomplish the goals and objectives.

NOTE: A simple goal for a confined-space accident can be to rescue the victim while protecting your personnel.

Some emergencies require that more than one agency be actively involved and at times the interests of those agencies may compete against one another. To avoid having competition between the different agencies, a **unified command** is recommended. Unified command occurs when the commanders of the different agencies work together and sort out their interests, as shown in Figure 5–5. Together the members of the unified command develop an action plan and direct resources so that the competition is minimal and safety and efficiency are of top priority. Examples of incidents when unified command would be appropriate include hazardous materials incidents, large-scale emergencies, and smaller emergencies in which a crime may have been committed. The type of command that is used will depend on the incident commander's decision to use single command or unified command. It is important to remember that other agencies consider their work as important as your agency's work, so cooperation is critical to your success or failure. You must consult with their representatives. Unified command does not mean that you have given up your turf or handed the emergency to someone else; it means that you have recognized the importance of other emergency response agencies in handling the emergency.

THE ACTION PLAN

An **action plan** must be developed for every emergency to which you respond. Action plans start by defining the goals and objectives you want to accomplish. The goals are fairly broad in scope and the objectives work to narrow the responsibilities. A simple goal for a confined-space accident can be to rescue the victim while protecting your personnel. The objectives would narrow the goal by identifying what you would need to do, including securing the scene, setting up ventilation for the victim, determining if the victim can be removed without entering the space, and removing the victim. Standard operating procedures or standard guidelines assist you with action planning in that they are established, routine actions. You, as the incident commander, will not have to start from scratch at each accident. Regardless of what is happening when you take command, your action plan must consider all activities and determine whether they are effective. Additionally, your action plan

FIGURE 5-5

Unified command is the type of command which results from the contribution of many different people or agencies.

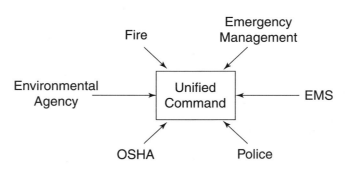

must maintain both unity of command and an effective span of control. If you cannot control the situation and the people involved, then you have no action plan.

COMMAND POST

► **command post**

the location from which all incident activities are directed by the incident commander.

The need to establish and use a **command post** will depend on the complexity of the incident. Incidents that are fairly simple and of short duration do not require a formal command post. They do, however, require you to establish command and you must command from a position that provides you with the best view of the emergency and that is accessible to the people who need to communicate with you. In short, do not establish a wandering position from which to command. Find a location and stay there unless the situation changes so dramatically that you either have to abandon the position or you can move in closer to a better command position, as shown in Figure 5-6. Larger-scale incidents demand a formal command post to be effective. A larger incident requires more people for command purposes and more agencies to handle the responsibilities. Establishing a formal command post identifies an area where command activities will be taking place. It allows you to separate yourself from some of the distractions of the emergency and to concentrate on managing the incident.

RESOURCE MANAGEMENT

► **resource management**

the allocation and maintenance of resources required during an emergency.

Resource management is a critical function of the incident commander. You will have resources that are not yet committed to the incident (your reserve), resources that are actively committed and in use, and resources that are no longer committed but unavailable because those resources (including people) have maintenance

FIGURE 5-6

When you establish a command post, identify it so that the location is obvious to everyone working at the emergency.

FIGURE 5-7

Establishing a staging area allows for management and coordination of resources. Any resources held in reserve will need to be put to work in an orderly fashion. That is one of the primary reasons for establishing a staging area.

Resource Status

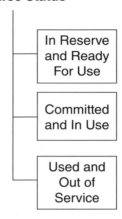

▶ **staging of resources**

a method of managing resources in which those not actively being used are kept or staged in a particular area in preparation for use.

needs, as shown in Figure 5-7. As the incident commander you must be aware of the status of your resources and you must try to stay ahead of the demands of the emergency. Do not count on resources that have been used at the emergency and have yet to be serviced for maintenance—people get tired, an apparatus runs out of fuel, and equipment breaks.

Any discussion of resource management must include **staging of resources.** Staging can be done in several ways. As people and equipment initially arrive at the scene of an emergency, the resources must be organized so that they can be accounted for, positioned where they do not interfere in emergency operations, and called up as needed. By establishing a staging area, you can keep your reserve resources in one location, ready and available. You may need a forward staging area that is closer to the emergency where selected resources are staged to support certain critical operations. This forward staging area might be necessary when it is not possible to get your apparatus close to the emergency, but pieces of equipment such as Sked™ stretchers, ventilation fans, spare SCBA bottles, and other items can be brought nearer to the scene. Regardless of whether you establish a single staging area or have a need to establish specialized staging areas, keep your resources ready and available. If arriving equipment is not kept in one area, you will waste valuable time looking for the people and equipment instead of putting them to work.

INCIDENT PRIORITIES

▶ **incident priorities**

the order of precedence given to the most basic goals of life safety, incident stabilization, and property conservation during an emergency operation.

The primary function of the incident commander is to assess the **incident priorities.** Although these priorities are simple, emergency responders can sometimes lose site of them. The basic incident priorities in order of importance are life safety, incident stabilization, and property conservation. You do not permit unreasonable life safety exposures to achieve incident stabilization. All emergency responders are familiar with different types of personal protective equipment designed to minimize their personal risk at an incident. Latex gloves, firefighter turnout gear, and police bullet-proof vests are examples of basic protective equipment worn daily. More complex emergencies can demand more complex protective equipment and procedures. Using the correct personal protective equipment begins to address life-safety command priorities.

COMMAND

Command is the only position that must be staffed at every emergency. As discussed, someone must always be in charge, but simply taking command does not end the incident commander's responsibilities. In addition to establishing command, the incident commander must set strategic goals and tactical objectives. Goals are broad statements of what the incident commander hopes to accomplish; for example, securing the scene and rescuing the victim. Objectives are more specific and they support accomplishing the goals. To rescue the victim the objectives might include ventilating the confined space, monitoring the atmosphere within the space, and then lowering a rescuer into it. All emergency operations must have goals and objectives, and the ability to effectively control an emergency will depend directly on establishing them and keeping the emergency forces working toward them.

From the goals and objectives, the incident commander will establish an action plan. The action plan involves defining which emergency responders will accomplish which parts of the goals and objectives. Your goal was to rescue the victim, the specific objectives included the need to ventilate the confined space, and now the action plan has the incident commander directing a person or persons to set up the fan in a specific place with the exhaust being directed to a specific area and so on. The more skilled the people who are performing the tasks needed to accomplish the action plan, the less direction they will need from the incident commander. Thus, you must take into account the span of control needed to effectively accomplish your action plan. People with higher levels of skills can be directed with simple statements: "Take that fan and vent the space" or "Vent the space by exhausting the gases to that area." People with lower skill levels may have to be directed as to where and how to get power to the fan, which way to face the exhaust, and other simple directions. Lower skill levels demand greater supervision, require longer times to accomplish objectives, and may require more resources.

Because the size of your emergency response team may expand to meet the needs of your incident, the incident commander must be prepared to develop a command structure to meet those demands. A simple incident will require a simple command structure, but as the emergency becomes more difficult to manage, the incident commander must meet those challenges by assigning others to assist in managing particular areas of the emergency. At a confined-space incident, a critical area that requires special attention is safety. The potential for many safety hazards exists and threatens the victims,

NOTE: All emergency operations must have goals and objectives, and the ability to effectively control an emergency will depend directly on establishing them and keeping the emergency forces working toward them.

NOTE: People with higher levels of skills can be directed with simple statements: "Take that fan and vent the space" or "Vent the space by exhausting the gases to that area."

emergency responders, and spectators. It will often be necessary to assign a person just to manage the safety issues at a confined-space emergency, and all persons involved in the situation must know this person's role and recognize his authority. Developing a command structure requires that needed incident management positions are anticipated, staffed, and recognized for their authority, as shown in Figure 5–8. This idea goes back not only to span of control, but also to unity of command, and the incident commander should only staff positions as needed. Developing and staffing an expanding command structure will require using more resources. Call for your needed resources early and be prepared to stretch your resources until the needed help arrives. A person inexperienced with using an ICS may fail to fill the needed positions or may fill every position and not have enough people available to actually perform work. There must be a balance between what you have available and what you need. Identify the incident command positions you need and fill them as soon as possible. Keep in mind that life safety takes priority over incident stabilization.

The whole purpose of an ICS is to properly manage the resources you need to accomplish your mission. Resources include people and equipment. If you expect that the incident will be long and drawn out, anticipate the need for relief crews, food, and basic physical needs for your people. If the incident has extensive equipment needs, anticipate where and how you will get the necessary equipment and how long it will take to get it to you. Regardless of whether your needs are for people or equipment, you must be aware of the status of the resources. The three basic states in which resources exist are those that are already committed, those that are in reserve for future commitment, and those that have been used and are no longer available for use. Emergency workers, like equipment, need maintenance. Expect that people will need water, food, and rest. Knowing the status of your resources will keep your operation flowing smoothly and help to avoid disaster, as emergencies tend to follow Murphy's Law.

SAFETY
Keep in mind that life safety takes priority over incident stabilization.

NOTE: The three basic states in which resources exist are those that are already committed, those that are in reserve for future commitment, and those that have been used and are no longer available for use.

FIGURE 5-8

In this particular incident (a drill), three positions of the incident command system have been staffed: Command, Safety and Operations.

SAFETY

Safety is a functional area that is involved in continuous discussion. The incident commander never relinquishes responsibility for safety, even if he assigns another person to actively take charge of safety, as shown in Figure 5-9. If the incident commander can effectively manage the incident and address safety with assistance, then all the better; but the reality of a confined-space incident demands that special attention be paid to the safety of everyone at the scene. For most emergency responders, a confined-space incident will be unique, it will not be routine, and the incident conditions present risks that are not always detectable without special equipment or knowledge. If the incident commander sees no other need to expand the incident command system, safety is the one area that requires an expansion and a take-charge person.

Coordination of the various agencies, resources, and activities is why you use an ICS. The incident commander is responsible for that coordination. Consider again the band which was discussed at the beginning of this chapter—without coordination they are nothing but a group of musicians. The incident commander has the responsibility to coordinate the available resources so that the end product is harmony.

PUBLIC INFORMATION OFFICER

Depending on the location and severity of the accident, the incident commander may have to be prepared to deal with the media. Although dealing with the press during an emergency may seem unnecessary and distracting, you need to realize that the press will get the information from one source or another, so it is best that they receive accurate and timely information from a qualified source. In the event that an incident is so severe that you will need to have the public take protective actions, such as staying indoors or evacuating

FIGURE 5-9

Here an individual has been assigned the role of Safety Officer. Safety is still the responsibility of the Incident Commander, but in this case, assigning a person to actively manage incident safety controls the risks associated with the incident.

NOTE: In the event that an incident is so severe that you will need to have the public take protective actions, such as staying indoors or evacuating from selected areas, the media can be a valuable asset to assist in getting out the proper information.

▶ **public information officer**

within the incident command system, the command staff position responsible for providing the press and media with information about the incident as authorized by the incident commander.

▶ **safety officer**

within the incident command system, the command staff position reponsible for incident safety including identifying potentially hazardous situations and the enforcement of safety procedures and safe practices.

▶ **liaison**

within the incident command system, the command staff position responsible for establishing and maintaining interaction with other agencies required to handle the incident.

from selected areas, the media can be a valuable asset to assist in getting out the proper information. The media will also be trying to learn the identity of the victim and the circumstances surrounding the emergency. The incident commander must control this information so that the victim's family is notified before the media reports the identity of the victim, and the facts concerning the circumstances need to be controlled to ensure accuracy. The incident commander must take responsibility for the release of public information and either present it directly to the press or assign someone as the **public information officer.** The press has a role to play in any newsworthy event—help them give an accurate presentation of the facts. At the very least, your reputation as an emergency response organization is at stake.

OTHER COMMAND SUPPORT STAFF

As both the size and complexity of an emergency grow, so do the demands on the incident commander. At a large-scale or serious incident, it may be necessary for the incident commander to assign people to fill the roles of a command support staff. Just as many top managers in business have a staff of support personnel, so too might an incident commander. Think of the amount of time the incident commander might be taken away from command if he has to directly deal with issues such as safety, liaison, and public information. By now you should realize that a confined-space incident creates significant and serious safety hazards to everyone present at the emergency. Except for the smallest incidents, the incident commander cannot hope to effectively carry out the duties of command and safety. Accept that as a fact and assign someone the role of **safety officer.** The safety officer reports directly to the incident commander and has the authority to stop unsafe acts.

It is important to have a good working relationship with other agencies. At a confined-space accident with serious injuries or death, you can expect that there may be other governmental agencies responding to fulfill their role. A state department of labor or OSHA representative may be required to participate in an investigation, especially if a death occurred. The medical examiner or district attorney's office may become involved. If a release of hazardous materials occurred or has the potential to occur, the state environmental agency or department of health may take over jurisdiction at some point. An incident commander trying to run an efficient emergency operation may not be able to confer with every agency at an emergency scene. The most efficient way to deal with a myriad of agencies is to assign an individual to serve as a **liaison** to them. The liaison will be able to listen to each agency and then sort out what information needs to be brought to the attention of the incident commander. Designating a person as the liaison officer will save the incident commander countless interruptions and provide the chance for each agency to be heard.

STAFFING OTHER FUNCTIONAL AREAS

Thus far we have discussed the topics of command and the command support staff. Four other general functional areas are outlined in the ICS. The incident commander holds direct responsibility for those functions until such time as responsibility is assigned to other people. The four other functional areas of the ICS are planning, operations, logistics, and finance and administration.

Planning is responsible for collecting, sorting, and interpreting data about the emergency. Collecting information is simple enough, but sorting through it and determining how it might affect emergency operations place special demands on this function. The functional area of planning should be staffed by a separate individual only if the demands for information gathering and interpreting exceed the commander's capability. The planning capability of the incident commander might be exceeded not only by the complexity of the incident, but also by the location of the emergency in relationship to the command post.

Operations is the functional area responsible for putting the action plan to work. Operations will direct individuals and companies to perform specific tasks. The most common reason to assign someone to be in charge of operations is to maintain a manageable span of control, as shown in Figure 5–10. There is no need to staff operations if the whole emergency response consists of an incident commander and four emergency team members.

Logistics is simply getting the resources you need when and where you need them. Assigning a person to bring a meter or tripod to the entrance of the confined space is not the same as assigning a person to take charge of the logistics function. The incident commander will remain responsible for this functional area until it is assigned to someone. Logistics only needs to be staffed when the demands for resources will exceed the ability of the incident commander to obtain those resources, as shown in Figure 5–11. An incident commander who is actively working on getting resources of a particular type or to a specific location will have to ask himself if the command of the incident is being compromised by the distraction of acquiring the needed resources. If the answer is yes, the incident commander will have to either assign a person to be in charge of logistics or assign someone else to be in command.

Finance and administration of an emergency is not a functional area most people think of as a function of the incident command system. After all, who thinks about what the emergency will cost or what paperwork needs to be filled out when trying to save someone's life. At some point, however, the cost of the incident will have to be tallied, so it is easier to keep track of your financial obligations and administrative needs during the emergency than to recall them afterward. Recording such simple items as the members who responded, what expendable items were used, and the name and address of any victims is part of finance and administration. It is done at most emergencies without anyone being assigned the role. Actually, the incident commander may already be doing the finance and administration part

NOTE: Planning is responsible for collecting, sorting, and interpreting data about the emergency.

NOTE: Operations is the functional area responsible for putting the action plan to work.

NOTE: Logistics is simply getting the resources you need when and where you need them.

FIGURE 5-10

Assigning an individual to manage the operations function of the incident is most often done when the span of control starts to become unmanageable for the Incident Commander.

FIGURE 5-11

This high angle rescue equipment is a good example of what can happen when no one is assigned to manage the logistics needs of an incident. There is a lot of equipment here and it is not sorted out. How do you know if you have what you need?

NOTE: Finance and administration of an emergency is not a functional area most people think of as a function of the incident command system.

of the job. Only if these requirements of the emergency demand special attention should you staff this functional position.

APPLYING THE INCIDENT COMMAND SYSTEM TO CONFINED-SPACE RESCUE

Now that you have become familiar with the incident command system, it is time to consider how to apply it during a confined-space emergency. Many people believe that the command system should be large and take advantage of as many positions as possible, which is not necessarily a bad idea, but the command system has to match not

NOTE: There is no sense staffing the command, operations, planning, and safety positions if you have only five emergency responders at the incident.

NOTE: The most logical and critical position to staff first is command.

NOTE: If you can only staff one other position at a confined-space emergency, then you must consider making safety that position.

only the demands of the incident but also the available resources (people). There is no sense staffing the command, operations, planning, and safety positions if you have only five emergency responders at the incident, which means that you have to prioritize. The most logical and critical position to staff first is command. Without a person to take charge of the emergency, set priorities, and develop an action plan there will be no coordinated action. So command must be staffed, and it must be staffed at every incident.

Then you determine your next need. Do you need a planning officer, a safety officer, or an operations officer? Well, what is the next most pressing problem that you face? Just as importantly, can you as the incident commander address that problem or do you need help? If you can address the problem, take care of it. If not, assign someone to take charge of it and assist you in managing the emergency. But what area do you think will be the most important? Safety, planning, operations, or some other problem? Stop and think about the incident, the nature of the problem, and the knowledge and experience of all the emergency responders. How many emergency responders have extensive knowledge and experience with a confined-space emergency? What about the attitudes of the emergency responders? Do they want to jump in and get the person out as soon as possible or are they cautious and calculating about what they are going to do? What hazards are obvious and what hazards are hidden at this emergency? Thus far you have been asked many questions—now let us develop an answer.

If you can only staff one other position at a confined-space emergency, then you must consider making safety that position. Safety can isolate the area of the emergency to keep out unneeded and untrained people. The safety officer can begin to characterize the atmospheric and physical hazards of the confined space, as well as the need to provide personal protective equipment, lockout/tagout, and other ways to manage the risks present. Safety can identify many of the limiting factors that will affect your operations and assist you—the incident commander—in addressing the incident priorities and developing an action plan.

The correct time to staff other functional areas depends on the situation. Suppose you get to a confined-space emergency and realize that it is beyond the capabilities of your organization. You have a mutual aid agreement with another fire department, police department, or EMS provider for such an incident. When you call this mutual aid group to the emergency you are still in command, but their level of training and experience is far beyond yours. What can you do to assist them? Well, you are still in command and you are still setting the priorities, goals, and objectives, but you will need to discuss what you want to accomplish with the person in charge of that group. Without giving up command, you can either establish a unified command or place the person in charge of the mutual aid group as the officer in charge of operations. If you place that person in charge of operations, you have expanded the incident command system in a logical manner and you have strengthened your ability to accomplish your goals and objectives. Even if you are trained and

NOTE: Remember, the whole purpose of the incident command system is to manage the incident and not have the incident manage you.

experienced enough to be able to effectively handle the emergency, you may still need to staff the operations position to maintain a manageable span of control. Remember, the whole purpose of the incident command system is to manage the incident and not have the incident manage you.

When deciding if and when to staff the other functional areas, you should anticipate the need based on the way that you expect the situation to progress. If you see that there are many unanswered questions or little information available regarding the emergency, you must determine if you can effectively command the incident while evaluating the situation. If you can do both effectively, do not staff the planning function; but if you expect a problem, get assistance by assigning responsibility to someone for the planning function. If you need additional resources or need to account for the costs associated with those resources, staff only those areas where you need help or those areas that can affect your ability to command the incident.

CASE STUDY

NOTE: To begin, establish command.

You are an officer in an emergency organization (fire, police, EMS) and you have responded with three subordinate emergency workers to a reported explosion in a manhole. Upon arrival you observe two burned utility company workers being loaded into the cab of a utility company truck. Before you leave your rig, the truck speeds away and a utility company supervisor is the only worker left on the scene. How would you implement the incident command system for this incident?

To begin, establish command. Someone must take charge of and be in control of this emergency. Because the level of knowledge you have about the incident is limited, you will have to develop more information before you can develop an action plan. You will have to secure the scene and find out if there are other victims that might still be at the scene. You also must keep spectators away and keep other emergency responders from entering the manhole before you find out what has happened. In securing the scene, looking for additional victims, and keeping emergency responders from entering the space, you have taken care of the life safety priority.

As you question the supervisor and investigate the scene for more victims you find out that only two people were in the space and they were able to escape on their own after the explosion. Because this manhole is fairly shallow and you can see the entire space without making entry, you are able to look into the space and see that there are no additional victims. Do you need to expand the ICS at this time and what should be your next priority?

Expansion of the ICS will depend in part on local protocols, but in this instance it appears to be unnecessary. Even if local protocols

FIGURE 5-12

This is a simple Incident Command System that has been set up for a simple incident. The Command function must always be staffed regardless of the size or complexity of the incident.

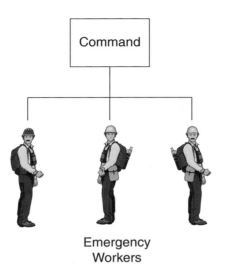

require that a superior officer be called to an incident of this type, you will only transfer command. You have an incident commander and there do not appear to be any other functional areas that need to be staffed at this time. Since your next priority will be incident stabilization, you should be able to effectively control the three other emergency workers at the scene and direct their activities, as shown in Figure 5-12. Your incident stabilization efforts will include determining what caused the explosion and monitoring the manhole to determine the presence of more combustible gases. There may also be a need to control ignition sources in the area and you may need assistance from other organizations for natural gas control, traffic control, investigation of the incident, and securing the scene. At this point you may turn the entire incident over to another government agency which will take charge.

The incident command system here was simple and straightforward. The only functional area that needed to have a person assigned to it was the role of incident commander. The other functional areas were not assigned because the incident commander could effectively handle those tasks. But what would you do about a similar emergency, with more victims or with trapped victims?

This incident is identical to the previous one except that upon arrival you are told that you have two victims in the confined space who are trapped 30 feet below. You must establish command, as always, and you must address the first priority of life safety, but you will need additional help. You have victims who may die within the confined space if they receive no assistance (they may even die if you provide assistance) and, as all emergency responders, your personnel are ready to take action. At this point you must prepare to expand the incident command system. You will need to assign an individual to take charge of safety and begin looking at the hazards of the space, the incident scene outside, and the hazards your personnel will face. Because there are only four emergency responders on the scene, you will have to call for help and also organize the emergency. As additional emergency workers arrive, you will have to control and direct their actions. If your call for help brings more people than you can hope to manage, you will have to assign people to maintain an effective span of control.

It is now time to think about assigning someone the job of operations officer. Depending on the level of skill needed for this rescue and the number of people required for it, you may even have to assist the operations officer by assigning people to control those who are being managed by him. The operations officer must also maintain an effective span of control and there should be people managing the rescue team, emergency medical services, and any other special needs to support operations. Span of control problems are the most common reason to staff the operations role and to break the operations function into smaller, more manageable units. Span of control is dependent on many features—not only the ratio of supervisors to subordinates—and one of those features must be the degree of risk in any given operation. The greater the risk, the smaller the span of control.

As you can see, the ICS needs of this emergency are expanding. Since you established command at the beginning, you have started a system that can expand with the emergency. Other needs may also have to be addressed, such as logistics, information for the media, coordination with other agencies, and planning, shown in Figure 5–13. With a complete incident command system in place, you will have the means to manage an emergency effectively.

| NOTE: The greater the risk, the smaller the span of control.

FIGURE 5-13

More complex incidents require more complex command structures to effectively handle the emergency. This structure has, hopefully, begun with Command being staffed and then expanded from there as required.

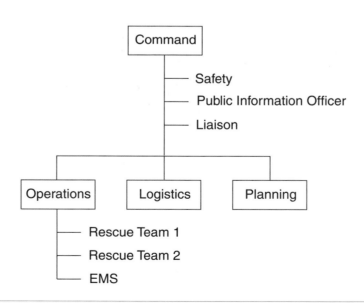

■ SUMMARY

Most if not all emergency responders know that someone must take charge during an emergency; however, this control and coordination will always come from a single source. Emergencies can be confusing and dangerous operations involving victims, emergency workers, and spectators. Not all of those dangers will come from the conditions found at the emergency. Some dangers will be created by the emergency workers themselves. The value of an incident command system then becomes crucial in that the risks that are present can be controlled, monitored, eliminated, or reduced. This organization of the site is only possible if the direction of the operations comes from one source and all personnel are working from the same action plan. Without coordinated and centralized direction, people will operate on their own and thus impact the operations of others, at times with the potential to cause injuries or death. Therefore, command must always be established upon arrival at an emergency and maintained for the duration of the incident.

Establishing command is not the end of the ICS. The incident commander must pay attention to safety, unity of command, and the span of control of emergency personnel to people supervising their actions. But the incident commander is still only one person with limited capability. When incident conditions require, the incident commander must be able to expand the ICS to meet the needs of the emergency. Certainly safety is one of the most important areas that must be addressed at a confined-space emergency. By using a model incident command system and by practicing with it, the incident commander should not be overwhelmed by any one problem and ignore other important areas that will influence the safety of people. Should the incident need special attention in the areas of operations, planning, logistics, or finance and administration, the incident commander will need to assign those roles to other people and have them take charge. The incident commander must know where people are, what they are doing, and if their actions are coordinated with the action plan. Command must also ensure that the plan is working and on schedule. Having a usable ICS and taking advantage of it improves the potential for effective rescue operations and minimizes the risks facing rescuers, victims, and other people at an emergency.

■ REVIEW QUESTIONS

1. Explain the role of command within the incident command system. What functional roles is command responsible for? How would command share that responsibility? When would it become necessary to expand the system?

2. Explain what is meant by span of control and how the degree of risk faced by rescue workers can affect it.

3. Why is unity of command necessary for a safe and effective emergency operation?

4. Using your experience as an emergency responder and your area of specialization (police, fire, EMS, etc.), identify one way in which you routinely address the priorities of life safety, incident stabilization, and property conservation.

5. To manage an emergency operation, the ICS is designed to address which of the following?
   a. Safety
   b. Unity of command
   c. Span of control
   d. All of the above

6. The ICS is considered an ____ ____ system.

7. During a major incident, it is not important to stage resources as equipment and people arrive. True or false?

8. The public information officer is an important function within the ICS. True or False?

9. Other functions of the ICS are
   a. Planning/operations
   b. Logistics
   c. Finance/administration
   d. All of the above
   e. Both a and c

10. The whole purpose of the ICS is to manage the incident and not have the incident manage you. True or False?

6 Strategic Rescue Factors

OBJECTIVES

After completing this chapter, the reader should be able to:

- recognize the value of preplanning as a benefit in rescue size-up.
- name the twelve strategic factors that can affect confined-space rescue operations.
- describe and explain the three incident priorities and how the order of implementation of the priorities can change with the incident conditions.

BASIC RESCUE SIZE-UP

Arriving at the scene of an emergency you look around and immediately begin to analyze the extent of the problem you are facing. Certain items are easily identified and you categorize the effect they will have on your operations. Other factors must be sought out and examined in more detail. As you complete your assessment of the scene, you realize that certain factors can be changed, but others cannot. The construction of a building on fire or the extent of injuries to a victim at a motor vehicle accident are two examples. Such factors are called strategic factors and they will determine your tactics for resolving the emergency. Strategic factors are so critical to the success of the operation that you must either work within the limits they set or you must work to change them. A confined-space rescue operation involves many strategic or limiting factors. It would be easy to devote volumes of any book to identifying all of these factors, but this book attempts to detail the most important ones and assist in identifying others.

The importance of preplanning cannot be overstated. You should not only preplan your response by recognizing the limits of your equipment and organization, but you should also preplan any needed assistance and identify other agencies that might be of value to you in a large-scale or difficult rescue. Preplanning should also include the different types of confined spaces in which you may possibly work. In preplanning the different types of spaces you should also take the time to review and train in those different spaces. If your agency is the designated rescue team for a particular facility, then you can require that facility to provide access to its confined space(s) for training. Of course, this requires cooperation between your agency and the facility, but the value of preplanning will make any potential rescue easier and faster when the limiting factors are known in advance.

The importance of different strategic factors will vary according to the incident, but you still need a basic set of factors to consider to begin your operations. When considering these factors, you will find that some will be obvious in the effect they have on your tactics, some will be easy to dismiss, and still others will require more intense review.

Questions about certain strategic factors can be answered by locating the confined-space entry permit. The permit will give you some sense of the type of work being performed, the hazards to expect, the number of people you should expect in the confined space, and other information that might be helpful during your size-up. But if there is no confined-space entry permit, you will be starting at ground zero and working your way out of the situation. The lack of a permit is not the fault of the rescuers and should not cause you to rush into a rescue. You must begin gathering all the information that a permit would contain and that is important to your rescue operation.

The following strategic factors will be detailed in this chapter:

- atmospheric hazards
- physical hazards
- location and accessibility of the confined space and the victim

> **NOTE:** If your agency is the designated rescue team for a particular facility, then you can require that facility to provide access to its confined space(s) for training.

- exposures
- construction
- contents
- available resources for rescue operations
- time
- special problems
- communications
- life
- weather

It almost seems as though atmospheric hazards should be an obvious consideration for confined-space rescue, but as you develop a checklist of items that need to be considered, do not take them for granted. By now you should be aware of the four potential atmospheric hazards of any confined space: toxic gases, flammable vapors, oxygen-enriched atmospheres, and oxygen-deficient atmospheres. To begin your size-up, assume that all potential hazards exist in the confined space. Only by carefully reviewing the situation and by monitoring the space should you dismiss any of the atmospheric hazards. If you do not have a flammable atmosphere and the oxygen level is between 19.5 and 23.5 percent, do not assume that you have no atmospheric hazard. Other gases could be present but undetectable with your equipment. Study Figure 6-1. Can you tell what was being done in the space at the time of the accident? Was a coating being applied which released toxic vapors? Is it possible that a tank was being cleaned and a layer of sludge was disturbed at the bottom of the tank thus releasing vapors? If you choose to vent the tank by using forced ventilation, where will the gases from the tank go as they leave? Will you be spreading the atmospheric hazard? You should also look at your own equipment as a potential source of contaminants. A portable generator with the exhaust being carried into the confined space can create its own hazardous atmosphere.

The potential for an atmospheric hazard requires that rescuers entering the confined space wear positive-pressure SCBA or a **positive-pressure supplied air respirator (SAR),** with escape bottle, until they can ascertain that no atmospheric hazard exists. Rescuers may have to wear respiratory protection for the duration of the emergency only to find out that it was unnecessary, but there is no harm in taking the extra steps when you are not sure of the existence of an atmospheric hazard. Consider the alternative if you were to downplay the atmospheric hazards.

What would you do, though, if your detection equipment showed a flammable atmosphere or an oxygen-enriched or oxygen-deficient atmosphere? What happened to create this situation? Was some unplanned work being performed in the confined space which created this problem? For an oxygen-deficient atmosphere, the SCBA or SAR would work, but can you do anything to assist the victim who is in that atmosphere? These questions bring us back to ventilation and where the exhaust gases will be going. You will want to direct the

▶ **positive-pressure supplied air respirator (SAR)**

a form of respiratory protection in which the self-contained air supply is remote from the wearer, the air is supplied to the wearer by means of an air hose, and the pressure within the facepiece is greater than the surrounding atmospheric pressure.

FIGURE 6-1

Depending upon the type of work being performed within the confined space, hazards may be introduced that you might not expect.

SAFETY

For an oxygen-enriched atmosphere you want to make sure that the exhaust gases from the confined space do not come into contact with internal combustion engines, flames, or other processes where they can cause or accelerate combustion.

► **location and accessibility**

as a strategic factor, the physical location of the confined space and the available means of gaining entry to it.

exhaust gases to an area where they will not create their own hazards. Flammable gases need to be diluted in air so that they are below the lower flammable limit, and you do not want to direct these gases to an area where they will come into contact with an ignition source. You also need to be concerned about any gases collecting in a low spot or an enclosed area where they can become a hazard to people there. For an oxygen-enriched atmosphere, you want to make sure that the exhaust gases from the confined space do not come into contact with internal combustion engines, flames, or other processes where they can cause or accelerate combustion. Materials that normally are difficult to ignite or do not burn readily in air can catch fire and burn rapidly in an oxygen-enriched atmosphere.

Physical hazards should take you back to the need for lockout/tagout. Are there sources of energy or equipment within the confined space that pose a hazard to the victim and rescuers? Physical hazards, however, go beyond lockout/tagout—slippery surfaces, sharp objects, and your own equipment in the confined space can be hazards. If you are able to access a confined space by using a ladder, but the ladder is unsecured and you fall from it, that too is a physical hazard. You should look not only in the space, but also outside of it to check for potential hazards. A hammer that is laying outside the opening of a confined space and is accidentally kicked in and hits you on the head does not hurt any less because it came from outside the space. Look for and control the physical hazards that are present, as shown in Figure 6-2.

The **location and accessibility** of both the confined space and the victim may be one of the most critical factors you will face. Ideally, all confined spaces would be at ground level with flat surfaces above and the victim would be visible and on a harness and retrieval line. Now for a real-world situation, your victim is at the bottom of a water tank on the roof of a five-story building and the tank has a curved roof, as shown in Figure 6-3. In this type of situation, preplanning will pay off. If you know the limits of the confined space and the access to it, you can prepare a plan in advance to gain access. Unfortunately we cannot preplan every emergency and we have to deal with each situation at hand. You should first consider if you have the necessary skills and equipment to handle this emergency. If not, then you need to either get assistance or develop an alternative action plan that allows you to

FIGURE 6-2

This trench has a variety of physical hazards surrounding it that can easily be knocked or dropped into the space and injure people.

NOTE: Rethink the basic philosophy: "How can I minimize the risk to operating personnel?"

work with existing skills and equipment. The time delay in getting other, more sophisticated rescuers to the scene or the time it will take to develop an alternative plan may result in the victim dying. Although none of us wants the victim to die, you must accept the fact that you were forced to operate within your abilities. Do not think that poorly planned and futile rescue efforts which injure or kill emergency responders would have made a difference in saving the victim.

We now return to a simpler rescue scenario—one in which your chances of success increase. If you arrive on the scene of a confined-space accident and your victim can be retrieved without anyone having to enter the area, then proceed with the rescue. Simple solutions that work are generally best for both you and the victim. You are looking to minimize both the amount of time it takes to rescue a person and the risk to which you expose your people. Consider the scenario, however, where your victim is not visible, such as in a sewer line with access through a manhole. Rethink the basic philosophy: "How can I minimize the risk to operating personnel?" You do not know where your victim is located or if you even have a victim. If you are going to put personnel into the manhole to look for a victim, then what can you do to reduce the risk? Certainly you want to secure your rescuer with a retrieval line and equipment, you want to monitor the atmosphere within the space, provide respiratory protection, and protect them against physical hazards. Then consider any potential limiting factors. Do you have the proper type of harnesses and can you set up your retrieval equipment? If yes, continue on to the other factors; if no, you will have to change your tactics. Does your atmospheric monitoring show any hazards? If yes, how

FIGURE 6-3

The confined spaces shown at the top of this building (the water towers) would provide a difficult rescue situation due to the limited access to them. Pre-planning this type of rescue would save time, increase safety, and provide for a more efficient operation.

▶ **exposures**

the people, property, and systems that may be affected by the confined-space rescue operations.

will you overcome those hazards? Will your respiratory protective equipment allow the rescuer to pass through the opening of the space without taking the equipment off, as shown in Figure 6-4? Are any physical hazards present and can you control them?

The process of sizing up an emergency presents many questions that need to be answered. Some questions will be easy to answer and some will take time and hard work. In fact, most experienced emergency workers will realize that one of the most difficult and frustrating parts of an emergency is getting accurate information about what has happened. The information may come to you in bits and pieces and you will have to assemble the big picture. Asking the right questions can save valuable time and lives.

Another factor pertaining to your size-up is identifying exposures. For firefighters, exposures usually mean those areas where the fire can spread. For our purposes, **exposures** refer to those areas where hazards, especially atmospheric hazards, from the confined space may spread. Actions that you take in attempting to enter the confined space to rescue the victim may easily carry out contaminants. In those cases when you are venting the area, make sure you know where your exhaust gases are going. Factors to consider include the location of the confined space (inside or outside of a building), elevation of the space and the potential for gases to drop or collect in low areas, the type of gases and the vapor density of the gases (if known), the limitations of your equipment (vent hose length,

FIGURE

People entering this confined space for rescue are able to enter while wearing SCBA. If the rescuers could not enter while wearing and using SCBA, how would you overcome that problem?

FIGURE

Level "B" chemical protective clothing is designed to provide protection from splash hazards. The SCBA is typically worn outside of the suit, and it would be possible to wear a Class II or Class III harness with this type of protective equipment.

amount of air movement, and power source for the equipment), and whether ventilation should be used.

Now these factors only cover atmospheric hazards, so how would you handle chemical contamination of equipment and personnel entering the space? Is there a need for specialized chemical protective clothing, as shown in Figure 6-5, and decontamination of the victim and rescuers? There was a successful confined-space rescue made in New Jersey where the rescuers learned after the rescue that the space was contaminated with PCBs. Not only were the victim and rescuers contaminated, but contaminants were also spread to fire apparatus, an ambulance, and a hospital emergency room. When you are faced with a potential exposure problem, work to minimize the extent of the exposure. Minimize the spread of contaminants and know where the contaminants will go and what they will do when they get there. If the resulting effects of the contaminants reaching an exposure are unacceptable, work on another solution.

Most of us would not ordinarily consider the **construction** of a confined space a strategic factor. Often the areas are steel, concrete, or some other substantial material, but what happens when the confined space is an old brick-lined manhole leading into an underground area, or a trench which has been covered for the night, as shown in Figure 6-6? Is the structure of the confined space solid enough and stable enough for you to work safely? Suppose the incident was the result of an explosion of flammable gases within the space. Then you have an area that may have structural damage or the opening into it may have been compromised. The construction of the confined space becomes important because your entire operation now depends on either the stability of the structure or your ability to make a new opening to gain entry. An explosion inside of an "empty" concrete and earth-bermed, natural gas storage tank killed all of the workers inside the tank. The fire department had to work with heavy equipment in removing and lifting the tank roof just to retrieve the bodies. Not considering the

▶ **construction**

the materials from which the confined space is built.

▶ **contents**

for confined-space rescues, the materials within the confined space. These contents may be either gas, liquid, or solid and may or may not contribute to the limiting factors affecting the confined-space emergency.

▶ **resources**

the people, supplies, and equipment required during an emergency.

other problems associated with this incident, would you have the resources in personnel and equipment to handle this situation?

Identifying the **contents** of the confined space helps you to realize its potential problems. The contents of a tank, vessel, or other confined space should lead you to some idea of the hazards you will have to face and resolve. If you pull up to a 250,000-gallon aboveground storage tank, as shown in Figure 6-7, and the witnesses at the scene tell you that the tank contained water for the plant's fire protection system, would that make you feel more comfortable than if they had said it contained acetone? Of course it would. Acetone is a flammable liquid with mild toxicity and you would have to be concerned about a flammable atmosphere, a toxic atmosphere, and a potentially oxygen-deficient atmosphere. The water should only present a potentially oxygen-deficient atmosphere, right? Wrong! That assumption may sound logical, but what was going on in the water tank when the accident happened? Could it be that algaecides were added to the water in the tank since it was basically standing water, or that nothing was added to the water but the tank was drained, repaired, sandblasted, and was being repainted? The contents went from water to the vapors from the paint. Different material and a different hazard are now present. The contents of the confined space are not only what was in the space originally, but may also be a result of the work being performed within the space.

Thus far in this chapter most of the discussion has been about confined spaces and the limitations they present. Your limitations, however, are also important. The **resources** you have available will affect your entire operation. To begin, consider how your people can be utilized at a confined-space incident. Their levels of training and experience must be high enough to meet the requirements of the emergency. Someone who received training in high-angle rescue a year ago, and has not practiced his or her skills, may not be able to perform well. How do your people react during an emergency—are

FIGURE **6-6**

This trench is a confined space because it is covered and may contain atmospheric hazards from being covered. However, it is still a trench, and you will still need to address any hazards that a trench presents in addition to the atmospheric hazards.

they level-headed and calm or do they tend to be alarmists? You need people who will support your operation, not panic and tear it apart. As a strategic factor, the level of training and experience of emergency personnel will have a direct bearing on the emergency operations, as shown in Figure 6-8. Members of a volunteer rescue unit may work at the facility and might be able to provide you detailed information about the incident you are facing. If you are a career department, you may have members who worked at the facility in the past.

In considering your resources used for rescue, you can have the best trained people, but if your equipment is unsatisfactory, you will have a hard time overcoming the deficiencies. Now this is not to say that your organization has to be prepared for every possible emergency. Ideally you prepare for the types of emergencies you expect on a regular basis and plan for the rest. Know each piece of equipment and its designated purpose. Work within the limits set by the manufacturers and the various standards for different types of equipment. To overcome deficiencies in your equipment, know in advance where you can get the additional things you will need. Preplan not only the confined spaces in which you may make a rescue, but also the resources you will need and how to obtain them.

Time is a strategic factor for many different reasons. The time of day during which a confined-space emergency occurs will affect the number and types of available personnel you have, the response time (e.g., traffic), and the amount of time between the actual confined-space accident, the discovery of the victim, and a call for help. Additionally, time of day will determine whether it is light or dark as well as changing conditions such as light, tem-

▶ **time**

as a strategic factor, the time of day, the day, week, or year and the relative impact that it will have on emergency operations.

FIGURE 6-7

This confined space can easily be taken for granted. After all, it is only a water tank, and what hazards can it contain when it is empty? Rescuers could easily overlook any hazards that might be present in this tank.

FIGURE 6-8

As a strategic factor, the level of training and experience of emergency personnel will have a direct bearing on the emergency operations.

perature, and wind speed and direction. Certain times of the year will bring climatic conditions, seasonal changes in traffic patterns, and the need for additional personnel and resources to cope with such conditions. When considering time, which many of us take for granted, at dusk where can you get needed lighting equipment and how long will it take to get to the scene and set up?

Any discussion of time would not be complete without considering **lead time.** Lead or reflex time is the amount of time it takes to accomplish a goal or objective, as shown in Figure 6-9. Because the incident commander has ordered something to be done does not mean that it will occur instantaneously. Not only does it take time for people to understand and begin to carry out an order, but also for the personnel and equipment to get in place and set to work. The actions of people and equipment, and the timing of each event, must be coordinated to avoid getting ahead or falling behind. If you gave an order to ventilate a space for 20 minutes before your rescue team entered, but the ventilation equipment was 15 minutes late in its operation, you would only have a 5-minute ventilation operation. Be aware of the timing that it takes to accomplish different objectives.

Special problems is a broad category and one that can be easily dismissed; however, this strategic factor is more like Murphy's Law: "If anything can go wrong, it will at the worst possible moment." If you have an incident during which special problems start to become limiting factors, you may even believe that Murphy was an optimist. Special problems can range from simple problems, such as having no electric power close to the confined space to run electrically powered lights and equipment, to more sophisticated problems, such as multiple, interconnected compartments that you must go through to get to a victim in a space. Special problems can be solved or at least identified through preplanning, as shown in Figure 6-10, but in the real world we know that preplanning is a best-case scenario. When you run into the unexpected and it becomes a limiting factor, you

▶ **lead time**

the amount of time that is required for a specific action to occur after notification is made of the need for that particular action.

▶ **special problems**

as a strategic factor, this is a broad category for problems which are unique to a particular incident and would only be likely to recur on an infrequent basis.

<figure>FIGURE 6-9

Planning, initiating, and coordinating rescue activities takes a certain amount of time known as "lead time." Be sure to include this lead time when implementing your action plan.</figure>

<figure>FIGURE 6-10

Pre-planning is a valuable tool for confined space rescue. These tanks are interconnected at the top with walkways. You can only access the tops of the center tanks by climbing onto the tanks on either end first.</figure>

► **communications**

the act of sending and receiving a message and having the message understood by the receiver.

will have to devote time and resources to resolving the problem. That can have a ripple effect on all of your other factors. Whatever you do to resolve the problem, always keep in mind the safety of your rescuers and your victim. Divide the problem into simple sections and address each section by solving the problems more easily. In this way you will build your simple solutions into a more complex operation.

Communications requires a message, a person to send the message, and another person to receive it. Not only must there be a sender and a receiver, but the message must also be understood after it has been received. The process sounds simple enough until you bring factors such as noise, anxiety, respiratory protective equipment, and communications equipment into the picture. Inside of a confined space there may be an echo which eliminates or interferes with communications or there may be physical barriers which take the rescue team out of sight of the incident commander or safety officer. At times, anxiety may be a factor in that people may filter a message to fit what they expect to hear or they may overreact based on a partial message. Respiratory protective equipment (SCBAs and SARs) can effectively eliminate most spoken communications without special equipment. Radios, although effective, are limited in a variety of ways, as shown in Figure 6-11. There may be interference from nearby electrical equipment, the construction of the confined space, the ability of the user to transmit a clear message or transmit at all, and the strength of the battery. Important questions to consider now are (1) Can you send, receive, and understand messages between the rescue team and the incident commander or safety officer (or both)? and (2) If you cannot communicate verbally, are nonverbal communications a realistic alternative? Consider what the rescue team must endure to communicate with other emergency personnel outside the confined space. The rescue team members may only be able to give yes or no answers to questions or they may only be able to transmit on the radio if they stop what they are doing, find and press the talk button on the radio, and then transmit the message. Communications must be preplanned—there is no other alternative. Not only must you have basic communications for the rescue operation, you must also have unmistakable emergency signals. The emergency may be inside or outside of the confined space, but everyone should be able to identify and understand what the emergency signal means.

FIGURE

At times, standard equipment used by emergency responders is difficult to use in a normal fashion. The radio shown here has been taped to the leg of the rescue entrant so that it can be used with the harness and ropes.

► **Life hazard**

as a strategic factor during emergency response, the threat posed to the victims, emergency responders, and spectators.

► **weather**

as a strategic factor, the effects that can be expected due to temperature, wind, precipitation, and other climatic factors.

SAFETY

Any consideration of weather must include the wind.

| NOTE: Life safety is always the first priority.

Life hazard as a strategic factor has three basic groups of people that need to be addressed: emergency responders, the victims, and the spectators, as shown in Figure 6-12. Identify people by the group of which they are a part and address the hazard presented to that group. Emergency responders who have no active assignment and are merely watching the incident are simply well-trained spectators.

Weather as a strategic factor is unique in that it may have no bearing on the incident or may be a critical factor. Certainly at a rescue where the victim is in a storm sewer and heavy rains are falling, weather will be a critical factor (and one which you can do little or nothing about). What about when the victim is indoors in a process vessel during heavy rain? Your victim may not be directly affected by the weather, but what about the availability of emergency units or the response time due to the weather? Suppose that it is snowing heavily and the confined space requires vertical entry. Will your people be able to safely work on the surface above the entry point or will it be too slippery? Weather will also affect your personnel in that they may be exposed to severe heat or cold.

Any consideration of weather must include the wind. You need to be aware of the direction in which the wind is blowing, as shown in Figure 6-13. Wind could possibly carry an atmospheric contaminant into the confined space. The wind can just as easily carry a contaminant out of the confined space and expose other people to it. Wind speed is also important in that it can assist in diluting any atmospheric contaminants if it is fairly high. At other times when the wind speed is low, contaminants may be concentrated in that one location. Wind direction can also change, especially for an operation that is fairly long in duration or for one near large bodies of water.

ESTABLISHING INCIDENT PRIORITIES

The three priorities at any emergency are life safety, incident stabilization, and property conservation. Life safety is always the first priority, and includes protecting the victims, the emergency responders, and the spectators. You can actively protect victims by rescuing

FIGURE 6-12

Potential life safety hazards can occur to three groups of people: victims, rescuers, and spectators.

FIGURE 6-13

It is easy to only focus on the confined space. The flag in the upper right center of this photo is an indicator of wind direction and speed. Additionally, there is a large body of water shown in the background. What effect can this have on your operation?

SAFETY
Spectators are best protected by removing them from any hazardous or potentially hazardous area. Get the spectators away and you keep them from becoming victims if the emergency expands unexpectedly.

NOTE: The second priority is always incident stabilization.

NOTE: Property conservation is the third priority, which means that if it is necessary to destroy or damage property to save a life or to control the incident, then it is an acceptable cost.

them immediately or by eliminating hazards to the victims and rescuers and then rescuing the victims. You cannot protect the victims by ignoring the hazards to rescuers. To protect the rescuers you must do those things that reduce, eliminate, or control the risks they face. Monitoring the atmosphere within the space (as shown in Figure 6-14), providing the proper personal protective equipment, placing the minimum number of rescuers into the confined space, and using an incident command system are all examples of how you can protect rescuers. Spectators are best protected by removing them from any hazardous or potentially hazardous area. Get the spectators away and you keep them from becoming victims if the emergency expands unexpectedly. There was a fire in a propane railcar, and it exploded during the incident. The explosion killed thirteen firefighters and one spectator and injured more than ninety other spectators.

The second priority is always incident stabilization. Incident stabilization consists of those activities that stop the growth of the emergency and allow you to begin operations to reduce the size of and then end the emergency. Incident stabilization activities include keeping untrained or unequipped rescuers from entering the confined space, venting the space to provide air for the victim while rescue equipment is set up, or accepting that a victim is dead and changing your operations from a rescue to a recovery. There are myriad other activities that can be performed to stabilize the incident, but none of them are done at the expense of the first priority—life safety.

Property conservation is the third priority, which means that if it is necessary to destroy or damage property to save a life or to control the incident, then it is an acceptable cost. It does not mean that you have a free hand to tear apart the confined space or facility without cause. Firefighters routinely cause damage to property to fight a fire or to make sure that they have found all of the hidden fire. This practice is acceptable and routine. Imagine if firefighters had to battle a fire in a beautiful, historic building with ornate stained glass windows. The firefighters need to break at least some of the windows to fight the fire—if they do not the whole building will be lost. To save as much of the building as possible, the tradeoff is that some

FIGURE **6-14**

You cannot control the hazards to protect rescuers and victims if you do not know the hazards within the confined space. Here, a rescuer is shown monitoring the atmosphere within the confined space.

of the windows will be lost. If no windows are broken, the whole building and all of the windows will be consumed by the fire. The same logic applies to you. Do whatever damage you must but do not forget what you are trying to accomplish. Because your resources are limited at an emergency, avoid squandering them by assigning work that accomplishes no purpose. Excessive property damage can be the result of well-intentioned hard work.

Now that we have identified the priorities for emergency operations—life safety, incident stabilization, and property conservation—consider the order of implementation. During an emergency, it is possible that you have addressed the issue of life safety and are actively working on incident stabilization. Does that mean that you will then ignore life safety? No, but your active priority is now incident stabilization. If a situation arises when someone places herself at risk, you must immediately address that risk and once again make life safety the active priority. At times it is possible to actively work on all three priorities simultaneously. Suppose you are called to a confined-space rescue and on arrival find that the victim is wearing retrieval equipment, is alert but complaining of a cut on the leg from a power tool in the space, and no contaminants are present there. In this case, you may be able to simply retrieve the victim without entering the space and package her outside the area. By not entering the space, you have established life safety, incident stabilization, and property conservation simultaneously. At other times the scenario may not be as simple and you will handle each priority in the order of its importance. The order of importance never changes, rather the order of implementation may change depending on the incident. One final caution regarding incident priorities. You may be actively engaged in incident stabilization or property conservation activities, but if something goes wrong, such as a flash fire or an injury to a rescuer, you will be forced to again work on the first priority of life safety. At this time, control of the emergency becomes even more critical and the incident commander must insist that people not take shortcuts that increase risk to unacceptable levels.

CASE STUDY

You have been called to an incident at a factory and on arrival are directed to an underground, concrete vault that is approximately 8 feet high, 10 feet wide, and 10 feet long, as shown in Figure 6-15. The opening into the space is a manhole cover on top of the vault which is 30 inches in diameter. The vault contains an underground water main and valves for the fire protection system. The vault has approximately 16 inches of water at the bottom. Inside of the space are three victims, two face down in the water and one sitting against the wall of the space. None of the victims are responsive to your calls. You also have two other victims who are police officers and are outside of the space. The police officers inform you that they received the original EMS call at 1:00 P.M. and it is now 1:20 P.M. The officers entered the space in an attempt to rescue the victims but had difficulty breathing and were helped out by employees of the plant. The plant safety manager assures you that the confined space is not a permit-required space.

Questions to Consider

1. Identify and explain the strategic factors that will affect your operation.

FIGURE

Illustration of the confined-space accident referred to in the case study.

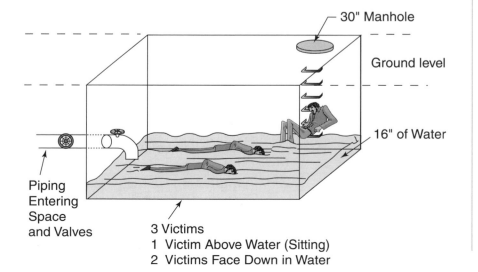

30" Manhole

Ground level

16" of Water

Piping Entering Space and Valves

3 Victims
1 Victim Above Water (Sitting)
2 Victims Face Down in Water

2. Identify the incident priorities and the order in which they must be implemented.

3. Would this space have been preplanned? Would pre-incident information influence your judgment?

Answers

1. The first strategic consideration in this instance is life. At the present time there are at least five victims. Two victims, the police officers, are outside of the space and three victims are inside. You are faced with at least one victim in the space who may be alive, although unconscious, and the immediate urge is to rescue that victim. The other two victims may be alive, but considering their position for the last 20 minutes or more, they may be dead. When you attempt to rescue the victims, the one who is sitting upright and out of the water should be removed first. He will have the best chance of survival. The rescuers will need to be protected from any hazard in the space if they are to be successful in their rescue attempts.

   Obviously, there are atmospheric hazards in this space, so it is essential to begin monitoring it immediately. At least an oxygen-deficient atmosphere must be suspected, and thus respiratory protection is a necessity at this emergency.

   The presence of water in this space is a physical hazard, but only in that there may be tripping hazards hidden below. The water is a severe hazard for two victims, and the one sitting must not be allowed to fall over and become submerged. No time can be wasted in entering the space to stabilize this emergency, but rescuers must be properly equipped for entry. Another physical hazard is the space, which is 8 feet deep, but this hazard can be controlled by providing proper access and using retrieval lines for the rescuers.

   The location of the victims is good. They are only 8 feet below grade and are not entangled. They must be lifted out of the space by some means other than brute strength unless they are small in size and can easily fit through the opening. These victims will have to be lifted using some device which will give the rescuers a mechanical advantage and will allow for each victim to be lifted completely through the opening. The victims cannot be easily carried up a ladder and through the 30-inch opening by a rescuer wearing SCBA.

   There appear to be no exposure factors at this incident because the area is outdoors. It would be expected that any atmospheric hazards exhausted from the space would not collect in another area. However, the actual contents of the space are hazards. The space is a non-permit confined space and thus is not expected to have hazards. The space involves piping and some valves, no known chemical contaminants, and only water below. In this particular case, the water was the problem. It contained an algae growth which consumed almost all the oxygen within the space and produced carbon dioxide gas. In fact, this non-permit confined space contained 7 percent oxygen and 3 percent carbon dioxide, all from the algae growth.

If you were to respond to this emergency, what resources would you have? How would you employ them? Would they be adequate to do the job?

The time of day is not an important factor, but rather the time that the victims have spent in the space. At least 20 minutes of submersion for the two victims is not good. As for the third victim, spending at least 20 minutes in an oxygen-deficient atmosphere is not beneficial either.

For this case study, there are no problems with weather, communications, or other special matters. It is easy to envision the problems that would occur with heavy rain or freezing weather.

2. The incident priorities are in order—life safety, incident stabilization, and property conservation. You must first worry about the safety of rescuers in order to address the life safety of the victims. After life safety is addressed, you can handle incident stabilization. In this case, the incident is fairly stable in that it is not getting any larger. Property conservation is a limited feature at this incident and may not need to be addressed at all.

3. Pre-incident information in this case would be of limited value. This space was not expected to contain any hazards, but it did. Knowing that the vault contained equipment for fire protection systems may lead the rescuers (if they are firefighters and know the facility from inspections) to assume that there are no hazards within the space and thus they would enter without a thorough size-up and without protective equipment. Preplanning is invaluable but is not possible at every confined space. A size-up takes time, but that time may mean the difference between a successful operation and the death or injury of a rescuer.

■ SUMMARY

When arriving at an emergency scene, you must find out as much information as you can about incident conditions. That information, when looked at in the proper context, will identify the factors that will affect your rescue operation. Some will not affect your operation, whereas others will force you to operate within the limits they set. You must identify the strategic or limiting factors and their influences, and adapt your operation to those limits. Keep in mind the need to identify and work on the incident priorities of life safety, incident stabilization, and property conservation.

■ REVIEW QUESTIONS

1. Identify at least two confined spaces within your community and describe at least four strategic factors. These confined spaces can be manholes, pits, silos, covered trenches, or any others in your area.

2. Based on your brief assessment of the confined spaces you identified in question 1, describe how the strategic factors may or may not be changed.

3. If you had the opportunity to preplan the confined spaces in question 1, what information would you look for?

4. If you are the designated team for a particular facility, you can request access to its confined spaces for training. True or False?

5. In a confined-space incident you should always consider that you have an IDLH atmosphere until you determine it to be less hazardous. True or False?

6. In an oxygen-enriched atmosphere, it is not important to worry about exhaust gases coming in contact with internal combustion engines. True or False?

7. What are the basic groups of people that need to be addressed during a confined-space incident?

   a. Emergency responders
   b. Victims
   c. Spectators
   d. All of the above
   e. None of the above

7 Ventilation and Inerting

OBJECTIVES

After completing this chapter, the reader should be able to:

- describe the benefits of ventilating a confined space
- recognize the types of equipment and methods of positive or negative pressure used for venting a confined space
- describe the value and importance of inlet and exhaust openings, duct hoses, and other features of ventilating equipment
- define the purpose of an inert atmosphere in a confined space

VENTILATION AND INERTING

Ventilation and inerting are two different topics, but both are intentional changes made to the atmospheric condition within a confined space. When venting a confined space, you are attempting to reduce or remove an atmospheric hazard. Inerting, on the other hand, is designed to reduce or remove the oxygen from a confined space so that the combustible vapors within the space do not have adequate oxygen for ignition. Inerting is never used during a confined-space rescue because the same lack of oxygen that will prevent ignition will also asphyxiate a victim. Inerting is mentioned in this chapter to make you more aware of the potential for a confined space to contain an inert atmosphere.

Ventilation

Ventilation of a confined space provides a victim who is breathing on his own an air supply which can have increased levels of oxygen and a reduced level of contamination. Thus, ventilation of the confined space is one of the first actions you take to protect the life of the victim and stabilize the incident. Before you ventilate you must consider several factors. To begin, you must have some idea of the atmospheric hazards within the confined space, as shown in Figure 7-1. You will need to know if the atmosphere is potentially flammable so that you do not have any ignition sources either inside or outside of the space where the gases will be vented. Toxic atmospheres that are vented from the confined space may come into contact with people outside of the space, so you will need to identify any toxic hazards as well as you can. If the oxygen content within the confined space is low enough, then that deficiency can also create an oxygen-deficient atmosphere outside the confined space. Defining the atmospheric hazard is one of the first steps to protect both you and any victims.

You have several considerations next, including the size of the space (as shown in Figure 7-2), the inlet location for fresh air and the

---

◐ SAFETY

Inerting is never used during a confined-space rescue because the same lack of oxygen that will prevent ignition will also asphyxiate a victim.

▶ **ventilation**

the systematic removal and replacement of air and gases within a space.

◐ SAFETY

Toxic atmospheres that are vented from the confined space may come into contact with people outside of the space, so you will need to identify any toxic hazards as well as you can.

FIGURE 7-1

You know that this space contains a hazardous atmosphere. Now you must determine what the hazard is.

CAUTION
CONFINED SPACE
HAZARDOUS ATMOSPHERE
ENTRY BY PERMIT ONLY

FIGURE 7-2

This confined space has a limited volume and a single, small (21″) opening for ventilation and access.

SAFETY
It would be better to have the air entering the tank as close as possible to the victim.

exhaust location for air leaving the confined space, the capabilities of your equipment, and the physical nature of the contaminants you are attempting to move.

The size of the confined space is a consideration because it will determine how much air you will have to move to exchange the air within the space. A 1,000 cubic feet (10′ × 10′ × 10′) space is fairly small. A fan or blower that moves 2,500 cubic feet per minute (cfm) will change the air 2.5 times per minute, which will create quite a breeze within the space but may also make it difficult to work because items may be blowing around the area. That same 2,500 cfm fan will be useless in a round storage tank that is 1,256,000 cubic feet in volume (a 200′ diameter × 40′ high tank), as shown in Figure 7-3. In fact, that 2,500 cfm fan will take at least 8 hours to change the atmosphere just one time. In a large-scale situation such as this, it would be better to have the air entering the tank as close as possible to the victim so that the victim receives the most benefit from the ventilation, as shown in Figure 7-4.

The location of the inlet and exhaust openings for air may seem fairly obvious, but the location of the exhaust is important for at least two reasons. First, air exiting the confined space may cause a hazard to people outside of the space. Second, the location of openings within the confined space can assist you in removing the gases depending on whether they are heavier or lighter than air. Gases that are heavier than air will tend to sink within a confined space and concentrate at the bottom of it. If both your inlet and exhaust openings are located at the top, you will have more difficulty removing gases at the bottom of the space. An area with two openings located remotely from each other would be the ideal for ventilation, as shown in Figure 7-5. You could use one opening for the inlet and the other for exhaust, but we all know that scenario is not always the case. When you are faced with a confined space that has only one opening, it will have to serve as both inlet and exhaust. When you use a single opening for inlet and exhaust, you use only a portion of the opening for the inlet and the remaining area for the exhaust; however, certain factors limit the effectiveness of this method. The first limitation you

FIGURE 7-3

This large, bulk storage tank will be difficult to ventilate due to the large volume of the space.

FIGURE 7-4

By blowing air into a space and placing the air hose near the victim, you can provide fresh air to the victim even if the space is large and difficult to exchange air in.

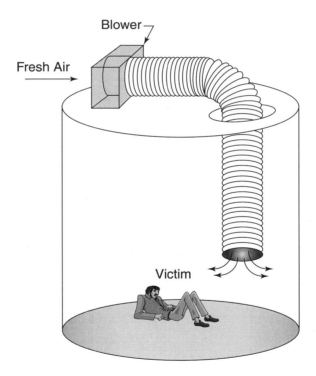

will have to overcome is the churning of air, as shown in Figure 7-6. If a fan is placed in an opening, the air being circulated by the fan can easily be blown out of the opening and back through the fan—the air is basically being circulated through the fan, into the space, out of the opening, and back through the fan. To prevent churning, use a hose attached to the fan to create a discharge point for the air that is remote from the opening. This hose should end at a point as remote from the fan as possible, shown in Figure 7-7. In fact, the ideal situation is to place the hose near the victim before any rescuers enter the space, thus providing the victim with fresh air and buying yourself time to set up and initiate the rescue operation.

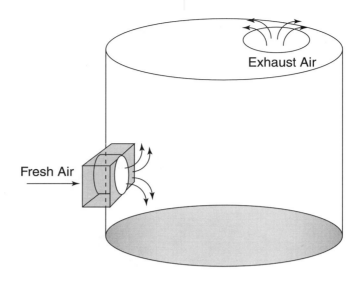

FIGURE 7-5

You must know where your exhaust gases are going as you vent a confined space. If these exhaust gases are heavier than air or the intake opening was downwind of the exhaust opening, what effect would that have on your operation?

Exhaust Air

Fresh Air

FIGURE 7-6

Churning occurs when air is blown through the fan or blower, enters and immediately exits the space, and is drawn right back through the fan. When you are venting a space you must be aware of how effective the air movement is. Air that is churning does not contribute to ventilating the space.

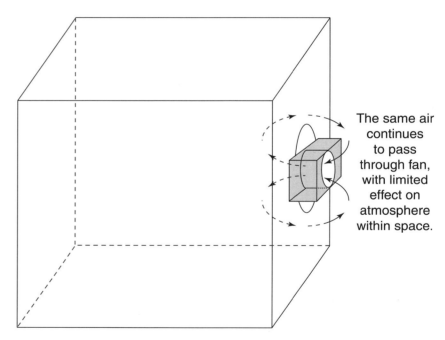

The same air continues to pass through fan, with limited effect on atmosphere within space.

SAFETY
When you are venting the space, you must know where the exhausted gases are going.

The location of the exhaust opening requires other considerations. As mentioned in chapter 6, the wind can be a consideration during a confined-space rescue. If the opening that you will be using for exhaust is facing into the wind or is upwind from your inlet opening, you may not be able to effectively move air through the space. Worse yet, the air you are drawing into the inlet may contain contaminated air carried by the wind from the exhaust opening. When you are venting the space, you must know where the exhausted gases are going. If you were venting a tank in a bermed or diked area and the exhaust gases were heavier than air, you would be taking the gases out of the confined space and venting them into the diked area which would effectively contain the gases. Now you have taken the hazard area and expanded it to outside of the space.

The type of ventilation equipment used and the capabilities of that equipment are two other concerns, as shown in Figure 7-8. How

FIGURE 7-7

When you have a space with a single opening in the space, it must be used for intake and exhaust. By using a hose to push air into the lower areas of the space shown, you will circulate air more effectively as it exhausts out of the top.

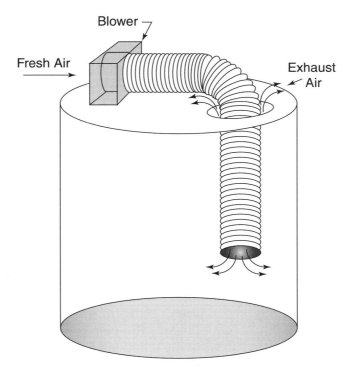

FIGURE 7-8

A gasoline powered blower and two electrically powered fans. Note the size and length of the ventilation hoses as well as the different sizes of the fans.

is the equipment powered? If the fan is driven by a gasoline or other internal combustion engine, are the exhaust gases being drawn into the fan and then pushed into the confined space? Does your fan use an electric motor to power the fan? Electric motors can be an ignition source and if you are using the fan to draw air out of the space and through the fan, you may inadvertently ignite flammable gases if the air and flammable gas mixture is within the flammable range. Many electric-powered fans are intrinsically safe when they are first purchased, but after being stored in a compartment on an emergency vehicle, used at other emergencies, and repaired by in-house mechanics, are they still intrinsically safe? Pneumatically powered blowers are available but require an air compressor for power. Some ventilation fans used by the fire service are powered by water. Except in unique circumstances, water-powered fans would be of little value at a confined-space emergency due to the large amounts of water required to continuously run the fan and the discharge of water from the fan motor. How many cubic feet of air can the fan(s) move? If you have three fans that can move 7,000 cfm each, you have the potential to move 21,000 cfm of air. However, can the fans be used in tandem (one behind the other), can they be stacked together, or does the size and location of the opening prevent you from using more than one fan at a time? You should also look at the attachments that you have for your fan. If you need a hose to direct air within the space, it must be compatible with your fan. You should also be aware that the length of hose will have an effect on how much air your fan can move. Passing air through a hose creates friction with the airflow. The longer the hose, the more the friction and the less air that you will be able to move.

When using mechanical ventilation equipment, should you blow air into the confined space (**positive-pressure ventilation**) or should you draw out the air (**negative-pressure ventilation**)? Let us look at each type of ventilation. Positive-pressure ventilation takes air from outside of the space and forces it inside, as shown in Figure 7-9. By forcing the air into the space, you can use a hose to direct the airflow so as to provide fresh air near the victim. When using positive-pressure ventilation you draw fresh air through the fan but do not contaminate it. In drawing fresh air through the fan you also avoid the problem of potentially flammable vapors coming into contact with the fan blades or motor. The motor of the fan is an obvious ignition source, but the fan blades are less obvious. Under certain conditions, the fan blades can create a potential static electricity hazard and ignite flammable vapors being drawn through the fan.

Positive-pressure ventilation is not the only type of ventilation you will use at a confined-space incident. Negative-pressure ventilation may also be useful. You may have to draw the contaminated air from the confined space to a location remote from the inlet opening. Your inlet opening may also be the one you will use for rescue and it may be so small that you cannot place a positive-pressure ventilation fan or hose into it. Using negative-pressure ventilation to assist positive-pressure ventilation is another consideration. By placing a positive-pressure fan at the inlet and a negative-pressure fan at the exhaust, as shown in Figure 7-10, you create a two-fan tandem and thus increase the airflow through the confined space. Neither the positive-pressure ventilation nor the negative-pressure ventilation is the ideal choice for all confined-space incidents. Each type has its advantages and positive-pressure ventilation often outweighs negative-pressure ventilation, but at certain times one type of ventilation will be a better choice than the other. That choice will depend on the circumstances of the incident, the capabilities of your equipment, and your own knowledge and experience. Ultimately, the three choices to mechanically ventilate a confined space are positive-pressure ventilation, negative-pressure ventilation, and a combination of the two.

▶ **positive-pressure ventilation**

the systematic removal of air, gases, and other airborne contaminants by using a fan to blow into a space to push the air or contaminants out.

▶ **negative-pressure ventilation**

the systematic removal of air, gases, and other airborne contaminants by using a fan to draw the air or contaminants out of the confined space.

NOTE: By placing a positive-pressure fan at the inlet and a negative-pressure fan at the exhaust, you create a two-fan tandem and thus increase the airflow through the confined space.

FIGURE 7-9

It is possible to use a single opening for positive-pressure ventilation. You must allow for a space at the top of the intake opening for exhaust gases to pass to the outside.

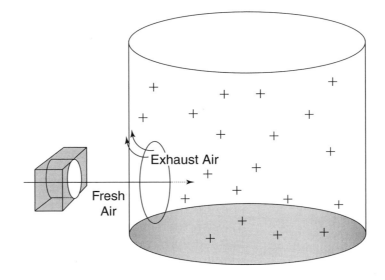

FIGURE 7-10

It is possible to use fans in combination for both positive-pressure and negative-pressure ventilation.

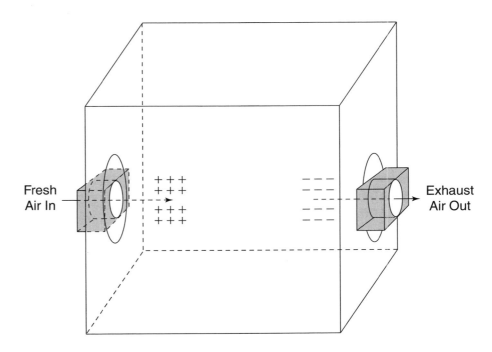

| NOTE: Ventilation also protects the rescuers.

| NOTE: Any failure of equipment must be considered a strategic factor.

If you are using fans or blowers to provide mechanical ventilation during a confined-space rescue, the loss of a fan or the effectiveness of the ventilation should be cause for concern. Using fans to ventilate a confined space not only protects the victims—ventilation also protects the rescuers. If the shutdown of the fan is the result of a simple problem, such as the plug being accidentally disconnected, fix the problem immediately and continue your operation. If the failure is caused by a more difficult problem, such as loss of power or a broken fan motor, you will have to consider if you can safely work around the problem or if it is necessary to pull the rescuers out of the space. Any failure of equipment must be considered a strategic factor. Your reaction to the problem will depend on the limits it places on the ability of the rescue team to safely continue operations.

▶ **Saddle Vent™**

a brand name for a ventilation device meant to be placed in the manway opening that will allow air to be directed through the opening with minimal obstruction to the manway for the movement of people.

New ventilation equipment is being developed specifically for confined spaces and any emergency response organization that is serious about developing a confined-space rescue team must be prepared to buy equipment designed for such use. In addition to fans and blowers, other devices include the **Saddle Vent™** (shown in Figure 7-11), heaters, and filters. The Saddle Vent™ resembles a flattened duct which takes air from an 8-inch blower hose, passes the air through the Saddle Vent™ (which is in the opening of the confined space) to another 8-inch hose, and only reduces the opening by 3 inches. Heaters are available to warm the air within a confined space and should be considered if you expect extreme cold, prolonged operations during cold weather, or hypothermia for rescuers or victims. These heaters may be a potential ignition source, however, and must be used with caution. There are also filters that can be placed in line to remove dust either from intake air or exhaust air as well as hoses in different lengths, carrying case for the hoses, and connectors.

Look at the location of your victim in regard to ventilation openings. A victim lying at the bottom of a confined space in an atmosphere which contains gases that are heavier than air is in the heaviest concentration of those gases. If the ventilation openings are low and near the concentrated vapors, it will be easier to remove or dilute those gases than if the ventilation opening were high and the gases had to be lifted out of the space. The closer the ventilation opening is to the gases in the space and the level of the victim, the easier it is to ventilate it and reduce the hazards to the victim and rescuers.

Inerting

▶ **inerting**

the introduction of an inert gas or a gas which will not support combustion into a tank or vessel so as to exclude oxygen from the tank or vessel.

For a fire to occur, three things must come together in the proper proportions: heat, fuel, and oxygen. If you want to prevent or extinguish a fire you either keep those three items separate from one another or you remove one of the three from the existing fire. **Inerting** is designed to remove the oxygen from a confined space. To displace the oxygen, an inert gas, or actually a gas that will not support combustion, is discharged into the space. These gases include nitrogen (as shown in Figure 7-12), carbon dioxide, argon, and others. The use of these gases is intentional and the gas may have been present in the confined space during normal process operations to prevent a fire or it may have been discharged into a confined space during a fire to extinguish it. Beside the control of a fire hazard, inerting may also be used to remove oxygen from a confined space, such as an enclosed storage bin, so that the product in the space does not decompose or deteriorate from contact with oxygen in the air. Regardless of why the atmosphere was inerted, you must determine, during your size-up, if an inerting system or gaseous fire-extinguishing system is connected to the confined space. Certainly preplanning would be an asset in the case of inerting and fire-extinguishing systems, but even with preplanning, you must be sure that the systems are locked out and tagged out. Without preplanning, you must determine if these systems are present and if they pose a problem. When you find inerting or fire-protection systems, have them locked out and tagged out unless that presents a more serious hazard to the rescue.

FIGURE 7-11

This is a Saddle Vent®, which is designed to allow the ventilation hose to be in place and in use while providing room for people to enter the confined space. (Courtesy of Air Systems International, Inc.)

FIGURE 7-12

The white tank, next to the two silos shown in this picture, is a liquid nitrogen tank for inerting the atmosphere within the silos. Without preplanning, you may not realize that this hazard is present.

In addition to gaseous fire-protection systems which displace the oxygen, you may find gaseous systems such as Halon™ type and dry chemical. The Halon™ type systems are supposed to be engineered for the specific location and most are designed to operate by interfering with the chemical chain reaction of the fire; however, they can still present a hazard. Although the systems are supposed to be engineered for the location and most of the gases are nontoxic at low concentrations, there are newer Halon™ replacement systems in service today which are more toxic and at lower levels. Other new systems are designed to reduce the oxygen content of the area to control the fire. Dry chemical fire-protection systems are designed to discharge a dry chemical fire-extinguishing agent. In addition to obscuring the vision of anyone within the space, these systems will provide an inhalation hazard to the victim who will not be wearing any respiratory protection. Once again, it would be better to lock out and tag out these systems to eliminate the problems of an accidental discharge.

Inerting systems and fire-protection systems which are designed to protect a confined space may be familiar to you. Regardless, you must be aware that they may be installed in the confined space where you have been called to make a rescue. Preplanning is the best way to become familiar with these systems, but for those instances when you have not had the chance to preplan, determine if the systems are present. When you find inerting and/or fire-protection systems, lock out and tag out the systems.

■ SUMMARY

Ventilation and inerting are intentional changes to the atmosphere within a confined space. Ventilation provides additional oxygen and can remove atmospheric contaminants there. When you are venting, the equipment must be capable of doing the job and it must not contribute to the hazard by becoming an ignition source or by pushing contaminants into other areas where people may be exposed. You must be aware of the need to vent the entire space, to know the vapor density of the gases within the space, and to provide air to the victim.

Inerting systems and fire-protection systems, on the other hand, are designed to control specific hazards within the space. On their own, these systems can introduce hazards to the rescuers and the victims. Control these systems so that they do not discharge into the space and endanger people there.

■ REVIEW QUESTIONS

1. What is the value of using a hose to direct air into a confined space?
2. What are some considerations that you must take into account when you locate your exhaust opening?
3. If you are attempting a rescue in a confined space that has a gaseous fire-protection system, would you lock out and tag out the system?
4. Describe what is meant by "churning" of the air when you are using a fan for ventilation.
5. During ventilation of a confined-space incident, people on the outside do not have to worry about the atmosphere near it. True or False?
6. There is nothing to be concerned about if you place your ventilation exhaust near a running engine. True or False?
7. Any failure of ventilation equipment would be considered a strategic factor. True or False?
8. _____ is designed to remove oxygen from a confined space.

C H A P T E R

8 | Safety

OBJECTIVES

After completing this chapter, the reader should be able to:

- describe heat-induced injuries, including: heat cramps, heat exhaustion, and heat stroke.
- recognize methods of preventing heat stress injuries.
- describe hypothermia and the temperature at which it can begin to occur.
- recognize the differences between Class I, Class II, and Class III harnesses.
- identify the hazards of noise within a confined space.

SAFETY CONSIDERATIONS FOR PERSONNEL

By now it should be fairly obvious that confined space rescues are not so much about the specifics of each rescue, but about the many different hazards that make them unique. So far, discussion has centered on atmospheric and physical hazards, OSHA requirements, duties of personnel, air monitoring, lockout/tagout, the incident command system, basic rescue size-up, and ventilation and inerting. For safety purposes, any time you find a hazard you should identify it, as shown in Figure 8-1, and then work to eliminate it. Unfortunately, that practice is not practical for all hazards present within a confined-space emergency. Some of the hazards you find will be easily and quickly resolved, whereas others will take so long to completely remove that you will have no chance of rescuing the victim. At times you will have to work within the specific limits set by the hazard and take extra steps to minimize the obstacle. Removing loose tools and equipment from around a vertical opening will reduce the chances of something being knocked into the space and striking someone, but it will not eliminate the need for head protection.

Even if you were able to control a hazardous atmosphere, how would you feel about having rescuers enter without respiratory protection? Your answer would depend on whether you could completely eliminate the hazard. When using mechanical ventilation the potential still exists for the hazardous atmosphere to reoccur within the space and you would still require respiratory protective equipment. But what if there was no respiratory hazard present to begin with? If your size-up of the emergency is able to confirm this and you do not expect such a problem to occur within the space, you would be inclined to eliminate the respiratory protection. So, doing away with the problem is the ideal, but during emergency operations you will have less than ideal conditions. At times the best you can hope

FIGURE 8-1

Given that this is a confined-space rescue scene, notice the numerous safety hazards present. These hazards include loose soil near the excavation, mechanical equipment close to the edge, no ladders for escape, and water in the space.

for is to reduce or control the hazard so that rescuers can enter and work safely with the personal protective equipment that is available. At other times when the available equipment is not adequate you will have to revise your action plan to accommodate this strategic factor.

Regardless of how you decide to handle the safety problems, keep in mind that the first priority of the incident is life safety—the life safety of the victims and rescuers. You must manage the risks to rescue the victim and to protect the rescuers. Weigh the risks you are taking against the benefits to be gained: When little is to be gained, then little should be risked. Emergency operations come with confusing and conflicting pieces of information. It is the incident commander's job (and everyone else's) to sort out the information. By keeping a balanced approach to the risks, safety should follow.

TEMPERATURE STRESS

Rescue operations may need to be conducted during times when air temperatures can be the cause of injury to rescuers. The cause of this high or low temperature problem may be from conditions within the space, as shown in Figure 8-2, or from ambient temperatures. Regardless of the cause of the air temperature, you must consider the effect it will have on rescue personnel and consequently on your rescue operation. A rescuer exposed to high or low temperatures will be affected in her ability to aid in the rescue; she may also become a victim in need of medical treatment or rescue. It is important to note that rescuers both inside and outside of the confined space can be affected by temperature.

The human body can efficiently work at ambient temperatures of up to 78°F. Above 78°F efficiency begins to decrease due to the effects of heat. Figuring in factors such as age, physical condition, sex, and protective equipment being worn, the effects of heat are increased. Most of us tend to think of heat as a problem when a "heat wave" occurs, but that is not always the case. Moderate air temperatures of 80°F combined with high humidity can cause the air temperature to feel the equivalent of 90°F and above. This is called the apparent temperature

SAFETY
Air temperatures can be the cause of injury to rescuers.

FIGURE 8-2

Confined spaces can be above or below the ambient temperature. Shown here is a steam coil for heating the product normally stored in this tank.

and the body will react as though the temperature really was 90°F or above. Add to this the other factors listed (age, physical condition, impervious clothing, etc.) and it is possible to cause heat stress injuries before the potential may be recognized. When temperatures are elevated, rescue personnel are exposed to the risk of heat cramps, heat exhaustion, or heat stroke. The best method of dealing with heat stress is to prevent the injury. Preventing heat stress means that you must reduce the impact that heat has on the human body by:

- cooling the atmosphere within the confined space
- minimizing the number of people exposed to the heat (including direct sunlight) by staging personnel in the shade or in an air-conditioned area, as shown in Figure 8-3
- setting up a fan to blow air across people to cool them
- preparing entry team members by ensuring that they drink liquids both before and after they have worked within the confined space
- minimizing the amount of time that personnel must wear impervious protective clothing, as shown in Figure 8-4
- having rescue personnel pace their work to match conditions
- medical monitoring when conditions dictate

Cooling the space may be practical—bringing fresh, clean, cooler air into the space will assist the victim and the rescuers. All limitations related to ventilation of the confined space apply. During the summer months the ambient temperatures both inside and outside of the space may be high and as such you would simply be moving heated air. You would have great difficulty cooling the atmosphere in a very large tank, and a limited number of entry and ventilation openings in any size tank may make air movement difficult. However, you may be able to provide a certain amount of cooling by setting up ventilation equipment so that air is blowing across the rescuers within the space. But, will that affect the victim and the safety of the atmosphere within the space? If the air current is too strong, airborne

FIGURE 8-3

Minimize the number of people who are needed to perform the rescue safely. Here is a small elevated platform crowded with rescuers.

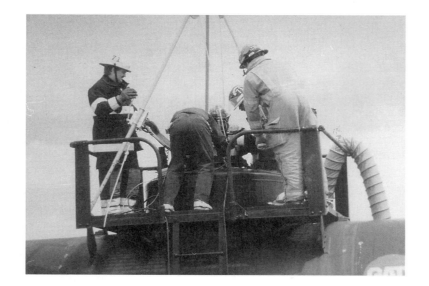

FIGURE 8-4

Protective clothing can increase the potential for heat stress injuries to rescuers. These level "B" suits can trap moisture from sweat and slow down the evaporative cooling effect.

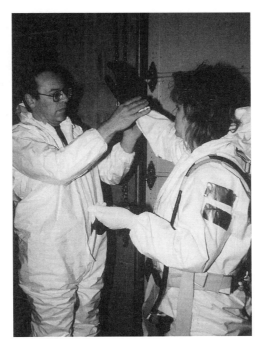

NOTE: Minimizing the number of people exposed to the heat sounds too simple to have to mention. Despite that simplicity, it is not unheard of for support personnel (the incident commander, standby personnel, support people working away from the incident site) to become victims of heat stress.

debris and dust may create more problems than it solves. An even simpler consideration may be that you only have one fan and you choose to use it for ventilation and for providing air to the victim.

Minimizing the number of people exposed to the heat sounds too simple to have to mention. Despite that simplicity, it is not unheard of for support personnel (the incident commander, standby personnel, support people working away from the incident site) to become victims of heat stress. Stage people out of the heat and out of direct sunlight even if this means setting up some type of shade covering. If shade is minimal or the temperatures are very high, consider setting up a fan to provide air movement. Sweating cools the body by providing evaporative cooling. Increasing the air movement will increase evaporation. Time spent idle being exposed to the heat will reduce the length of time people can work in the heat. Reduce the effects of heat and maximize the amount of time you will be able to work by staying out of the heat until you are needed.

Make sure that your entry team members drink liquids. Sweating reduces the volume of liquids within the body and you must replace those liquids, as shown in Figure 8-5. If you reduce the volume of liquids too much, the body will seek to protect itself and heat cramps, heat exhaustion, and heat stroke can follow. When the potential exists for heat stress injuries, the entry team members should drink a minimum of 12 to 16 ounces of water or a sports drink that can replace electrolytes, both before and after entry. Remember, it is not only the actual temperature, but also the apparent temperature, that can cause heat injuries. It takes about 20 minutes for liquids to be available for use by the body. If you drink liquids before entry you will have those liquids available while you are working. After the rescuers have left the confined space, they must restore lost liquids even if they state that they are not thirsty, because thirst may not be a reliable indicator.

SAFETY
Having rescue workers pace their work to match conditions at an emergency is always a good idea regardless of the temperature.

SAFETY
Do not discourage efficiency, but rather monitor people so that they do not rush or take shortcuts that may injure themselves or others or affect the rescue operation.

► **Medical monitoring**

basic medical evaluation of emergency response personnel by determining and recording basic vital signs such as blood pressure, respirations per minute, and pulse.

Most emergency personnel wear response clothing that holds in heat. Firefighters' protective clothing is designed to keep out both heat and steam from a fire. The clothing is also effective at keeping in body heat and preventing sweat from evaporating. Other emergency service personnel wear clothing that is designed to keep out the rain or cold and that is impervious, just like firefighters' protective clothing. Consider what the hazards are within the confined space and dress appropriately. Wear protective clothing when needed, but consider the effect that the apparel will have on personnel. When protective clothing is required and temperatures are high, you will have to accept that people's efficiency will decrease, that it will take longer to accomplish the assigned tasks, and that it will take more people to accomplish those tasks. For rescue personnel who are outside of any hazard area, consider if it is in their best interests to open up or remove the protective clothing while they are staged and awaiting assignments.

Having rescue workers pace their work to match conditions at an emergency is always a good idea regardless of the temperature. If you are the incident commander or other supervisor during a rescue, assign work to people that can be accomplished in a realistic time frame based on the working conditions. People must also work at the pace that you assigned them. Do not discourage efficiency, but rather monitor people so that they do not rush or take shortcuts that may injure themselves or others or affect the rescue operation. Pacing your work will allow you to accomplish the goals and objectives you need to complete, reduce injuries, and provide for a better flow of the emergency in that people will be able to maintain a level of control and thus succeed.

Medical monitoring of rescue personnel is valuable in that it can provide for early intervention of a developing medical emergency. It does not replace managing the risks at a rescue. You should have emergency medical personnel on the scene to provide assistance for the victim, but do you also need them for rescue person-

FIGURE 8-5

As you sweat you lose fluids. You must replace those fluids in order to help prevent heat-related injuries.

Replace Lost Fluids

nel? Do you have any specific hazards that may endanger emergency personnel and can those risks be best addressed through medical monitoring? A variety of reasons indicate the need for medical monitoring, and heat stress is certainly one of those areas that requires it both before entry into the confined space and after exiting the space. Vital signs such as pulse, blood pressure, respirations, and body temperature can indicate the presence of heat injuries. Having medical monitoring available reduces the risk of heat injuries by identifying what is going on within the body. A rescue team (entry team) member who has had an increase in body temperature from pre-entry monitoring to post-entry monitoring may need medical attention to prevent further heat injuries. Prevention of the heat stress injury or additional injury is the key factor.

In addition to preventing heat stress injuries, you must be able to recognize the signs and symptoms of heat stress, as shown in Figure 8-6. Heat cramps are indicated by severe muscle cramps in the abdomen and legs and are believed to be caused by loss of fluids due to heavy sweating. Heat exhaustion can cause the victim to feel weak, nauseous, dizzy, faint, and to have pale and clammy skin due to blood flowing to the skin and away from vital organs in an attempt to cool the body. Victims of heat exhaustion usually perspire heavily and the body temperature is normal or below normal. It is rare for a heat exhaustion victim to lose consciousness, but it can occur. Left untreated, heat exhaustion can lead to heat stroke. Heat stroke is the most serious heat injury and is considered a true medical emergency. Victims must receive rapid medical attention. It has been reported that up to 20 percent of the victims of heat stroke die. The signs and symptoms

FIGURE 8-6

When working in conditions that can cause heat-related injuries, you must recognize the signs and symptoms of heat cramps, heat exhaustion, and heat stroke.

Signs and Symptoms of Heat-Related Injuries

Heat Cramps

Muscle cramps - especially legs and abdomen
Normal body temperature and moist skin
Can advance to other more serious heat-related injuries

Heat Exhaustion

Headache, nausea, dizziness
Exhaustion
Normal or below normal body temperature and cool, moist skin
Can advance to heat stroke

Heat Stroke

May be unconscious
Hot, dry, red skin
High body temperature
Can lead to convulsions, coma, and death

of heat stroke include lack of sweating (even though the skin may be wet from earlier perspiration); dry, hot, red skin; elevated body temperature; and possible convulsions or unconsciousness. Heat stroke is caused by the failure of the body's system to regulate temperature, which may rise to 104°F or more.

Cold temperature injury is another problem of which you must be aware, even if it is less serious than heat injury. You may face a hazard from low temperatures due to weather conditions or due to the fact that the confined space was refrigerated. At low temperatures the body attempts to produce heat to keep up with the loss of heat to the cold. If the body is having difficulty producing heat, the blood supply retreats to the head and torso in an attempt to protect the brain and vital organs, which can lead to shivering and eventually loss of muscle coordination. If the condition is serious enough it can lead to **hypothermia,** which is a lowered body core temperature. Severe hypothermia can lead to unconsciousness and death. Hypothermia can occur at temperatures as high as 50°F if the person is wearing wet clothing.

Precautions for cold temperatures would include dressing for the cold to prevent loss of body heat, avoiding getting wet either from sweat or liquids which are in the confined space, and pacing your work. Recognizing the existence of a cold hazard may be difficult during warm weather. Preplanning a confined space would help, but failing that, you would have to discover the cold hazard during your size-up. Additional layers of clothing limit mobility and dexterity, and may be considered as much a strategic factor as heat. Prevention of the injury is the goal for cold temperatures just as much as it is for high temperatures. The effects of wind are a well-known phenomenon in colder climates in that weather reports typically contain windchill factors. Windchill is caused by colder air circulating around the body and drawing off heat, as shown in Figure 8-7. We typically think of that wind source as the naturally occurring wind in the environment, but certain processes such as mechanical ventilation produce a windchill effect. For people from warmer climates who are not accustomed to windchill, they may not recognize it as such if the equipment is the source. A ventilation fan that is moving air at 15 miles per hour is the same as a 15-mile-per-hour wind. At 40°F a 15 mph wind produces a windchill equal to 22°F. Precautions for prevention of cold injuries parallel those for heat injuries, with some changes, and include the following:

- heating the atmosphere within the confined space
- minimizing the number of people exposed to the cold
- minimizing the amount of time that personnel must wear impervious protective clothing, in this case to prevent sweating
- having rescue personnel pace their work to match conditions
- medical monitoring when conditions dictate

In attempting to heat the atmosphere within a confined space, you must consider whether it is safe to do so. The heater may be an

► **hypothermia**

lowering of the body's core temperature.

FIGURE 8-7

Windchill chart. Notice how the wind can create a cooling effect at mild temperatures.

Windchill Factor

Actual Temp (F degrees)	Wind (miles per hour)							
Calm	5	10	15	20	25	30	35	40
50	48	40	36	32	30	28	27	26
40	37	28	22	18	16	13	11	10
30	27	16	9	4	0	-2	-4	-6
20	16	4	-5	-10	-15	-18	-20	-21
10	6	-9	-18	-25	-29	-33	-35	-37
0	-5	-21	-36	-39	-44	-48	-49	-53
-10	-15	-33	-45	-53	-59	-63	-67	-69
-20	-26	-46	-58	-67	-74	-79	-82	-85
-30	-36	-58	-72	-82	-87	-94	-98	-102

ignition source; but the heat introduced into the space may also raise the temperature of materials within the space and cause greater vaporization. These vapors may be toxic, flammable, or both and may displace the oxygen within the space. Sometimes these conditions will not occur and it then may be in the victim's best interest to increase the temperature of the confined space. As with preventing heat injuries by changing the air within the space, be realistic about what you can accomplish.

Minimizing the number of people exposed to the cold means getting people out of the wind and any precipitation that is occurring (your heat loss is increased tenfold when you are wet), and into heated areas, and frequently relieving people who must remain in the cold to perform their assigned work. Both cold temperatures and hot temperatures may require additional personnel at the emergency scene for everyone's protection.

Wearing impervious clothing such as firefighters' protective clothing or Tyvek™ suits can lead to sweating. Sweating is designed to provide cooling and people who are exposed to the cold and are sweating heavily under their protective clothing can rapidly lose body heat. If you cannot prevent the sweating, then limit its effects by getting people out of the cold quickly.

As discussed, pacing your work is always a good idea. In cold temperatures you must pace your work because of the layers of clothing you are wearing and because of the effect the clothing has on movement and dexterity. Pacing your work during cold weather also paces your rate of breathing. Rapid breathing during cold weather can lead to an increased loss of body heat through the lungs and cooling of the blood. Pace your work to match conditions and accept that you are doing the best that you can under the conditions you are facing.

The use of medical monitoring during cold temperature conditions is designed to recognize and prevent hypothermia. People who become victims of hypothermia may not recognize it and may be prone to deny that they are suffering from it. Hypothermia begins when the body core temperature is reduced from 98.6°F to 96°F. Medical monitoring of vital signs, especially body temperature, will lead to proper treatment and prevention of more serious injury.

PERSONAL PROTECTIVE EQUIPMENT

Personal protective equipment includes that for the head, hands, legs, feet, body, respiratory system, and arms. A helmet or a hard hat is worn for head protection; however, firefighter helmets present certain problems. They are intended to provide protection during fires, which means that they are designed to shed water and keep heat and debris off the firefighter's head and shoulders. Therefore, the helmet is very wide from side to side and front to back, which makes it difficult to wear during entry into a confined space. A better helmet would be a caving helmet or a hard hat with a strap to hold it in place, as shown in Figure 8-8.

Respiratory protection must be provided by using either a self-contained breathing apparatus or a supplied air respirator. The two important factors here are that the respiratory protection must have its own air supply and that the equipment must not interfere either with entry into the space or with exit from it, as shown in Figure 8-9. Both pieces of equipment have limitations. A supplied air respirator (SAR), as shown in Figure 8-10, must have an escape bottle attached to it, typically of 5 or 10 minutes' duration. Additionally, specific limitations to the length of the air supply hose and the pressure within the unit must be regulated to ensure that the user is getting the proper flow of air. The air hose for the SAR may be no longer than 300 feet because longer hoses have too much resistance to the passage of air to provide the proper positive pressure and liter flow per minute to the facepiece. Another problem with using an SAR is that you will

| NOTE: Firefighter helmets present certain problems.

FIGURE 8-8

Shown here (left to right) are a firefighter's helmet, a rock climbing helmet and a hard hat with a strap. Notice the size of the helmets in relation to each other. If you expect rescuers to wear helmets, the helmet must be able to fit into the space and be worn without interfering with movement.

have approximately 300 feet of air hose to drag around and negotiate through obstacles within the confined space. However, SARs may still be your first choice for certain rescues. Supplied air respirators with an escape bottle have an almost unlimited supply of air available and the harness and escape bottle have a very low profile. This setup makes them ideal for entering tight openings. An SCBA has limitations in that the air supply is of a much shorter, finite duration (30-minute, 45-minute, and 60-minute rated), and the size of the SCBA can make it difficult or impossible to enter the confined space when there is a limited opening and rescuers are wearing SCBAs. You

FIGURE 8-9

Because of this person's body size he is able to enter this confined space wearing and using an SCBA.

FIGURE 8-10

A supplied air respirator showing the air supply, air hose, escape bottle, and facepiece.

▶ **National Fire Protection Association (NFPA)**

an international nonprofit organization advocating scientifically based consensus codes and standards, research, and education for fire and related safety issues.

▶ **Class I harnesses**

harnesses designed to support a one-person load for escape purposes with the harness fastening around the waist and either under the buttocks or around the legs.

▶ **Class II harnesses**

harnesses designed to support a two-person load for rescue purposes with the harness fastening around the waist and either around the thighs or under the buttocks.

▶ **Class III harnesses**

harnesses designed to support a two-person load for rescue purposes with the harness fastening around the waist, either around the thighs or under the buttocks, and over the shoulders to protect against inversion.

may conclude that all you have to do is temporarily remove the SCBA from your back while still wearing the facepiece and breathe off the unit. This conclusion is dangerously wrong.

Removing the SCBA to fit through the opening of the confined space places the wearer at risk in the event that she drops the SCBA and the facepiece is pulled off of her face. It might not seem dangerous if it occurs when she first enters the space, but what will happen as she negotiates her way around other openings and restrictions within the space? If you drop your SCBA and the facepiece is pulled off while you are in the space, you could very well be taking your last breath. Thus, respiratory protection has two basic types—SAR and SCBA—and both have limitations. If you only have SCBAs available to your rescue team, do not try to adapt them to an incident that requires SARs. Accept that the equipment you have and the way that you can use it is a critical factor—either change the conditions or change your tactics.

If you intend to be serious about confined-space rescue, you must have or have available for use the proper types of equipment. If frequent occasions in confined-space rescue warrant it, consider purchasing supplied air respirators, escape bottles, and any other equipment for SAR use for the rescue team. Although the rescue team will use SCBA, do not limit your operations to only those incidents when SCBA will work, without removing it for access. If you must take off your SCBA to enter the space and you become a victim, then how do you expect to be removed without removing the SCBA? If you are a rescuer who is now a victim, who will remove and hold your SCBA inside the space so you can be pulled out? Either fit through the opening with the SCBA in place and in operation (this is where smaller people become more valuable) or do not enter the space!

Retrieval equipment must be worn by rescuers entering the confined space, including the most basic piece of equipment—a harness. The **National Fire Protection Association (NFPA)** has a standard for rope (NFPA 1983 Life Safety Rope) which also addresses harnesses, which are classified as Class I, Class II, or Class III harnesses. **Class I harnesses** are designed to fasten around the waist and thighs or around the waist and under the buttocks and support a one-person load. **Class II harnesses** fasten around the waist and the thighs and are designed for rescue use when a two-person load may be encountered. Use of Class II harnesses involves certain limitations because they are not designed to hold you in the harness in the event you flip upside down (invert). **Class III harnesses,** as shown in Figure 8-11, are similar to Class II harnesses, but with the added advantage of straps which go over the shoulders and provide protection from slipping out of the harness in the event that the rescuer is inverted by accident. Class III harnesses are the choice for confined-space rescue not only because of the potential for inversion, but also because they can be made with additional attachment points that allow the rescuer to be lowered more easily. The additional attachment points can be located at the chest, upper back, shoulders, and the sides of the waist if needed. For restricted openings rescuers can

FIGURE 8-11

This is a class III harness with "D" rings at the shoulders and center of the back.

SAFETY
Class III harnesses are the choice for confined-space rescue not only because of the potential for inversion, but also because they can be made with additional attachment points that allow the rescuer to be lowered more easily.

▶ **American National Standards Institute (ANSI)**

an organization that administers and coordinates a voluntary private sector standardization system. Standards developed under ANSI are consensus standards created by representatives from various interest groups.

keep their hands in a position that provides the lowest profile and still be in an upright and stable position.

The selection of harnesses is often made on personal preferences and how they are constructed and how they feel while being worn. This practice is acceptable as long as the harnesses are designed and manufactured to meet a particular standard. By demanding that your equipment meets the standard, you can ensure that any equipment you purchase will perform as expected and will be designed for the intended use. Various organizations create standards, including the National Fire Protection Association (NFPA), the Occupational Health and Safety Administration (OSHA), and the **American National Standards Institute (ANSI)**. These organizations do not certify that any equipment meets a particular standard or criterion, but the standard contains information used by an independent testing laboratory to certify that a representative sample of the equipment has been tested according to the standard and meets the requirements. You will often find similar standards from different organizations for the same equipment. Organizations such as NFPA and ANSI maintain corresponding standards with one another, but it is important that you check the standard to determine who the intended user is for the equipment designed to that standard. SCBAs that are designed to meet NIOSH/MSHA requirements (OSHA's SCBA standard) provide a different liter flow per minute to the wearer than SCBAs designed to meet NFPA's standard. NFPA's standard is based on use by firefighters and the demands of their work, whereas OSHA is designed for general industry use and less physically demanding work.

Standards must also be current. NFPA routinely updates its standards every five years, and OSHA has some standards which have been in place, unchanged, since the late 1970s. By using the correct standard, you can purchase and maintain your equipment in a reliable fashion, you can expect the equipment to work as designed, and you can match the price of comparable equipment. Obtaining the

SAFETY
Wear a pair of leather outer gloves to protect the hands.

SAFETY
Consider if it is necessary to wear latex gloves while performing confined-space rescues.

SAFETY
The outer gloves must be discarded because of possible contamination.

SAFETY
Steel-toed work boots with soles designed to provide traction on most surfaces are ideal.

right equipment for the job according to the accepted standards helps to eliminate the guesswork involved.

Wear a pair of leather outer gloves to protect the hands. The gloves should be worn when grasping ropes to protect against rope burns. Leather gloves may not provide enough protection against certain sharp objects that may puncture or cut the gloves. Firefighter gloves designed to the NFPA Standard 1973 are difficult to use in confined-space rescue because they are generally bulky and do not provide for a good grip when handling ropes and cables. These gloves are designed for firefighting, so they must keep out heat and water and provide puncture and cut resistance. Confined-space rescue team members should wear gloves that allow them full use of the equipment and still provide adequate protection from the hazards, as shown in Figure 8-12. While on the discussion of gloves, consider if it is necessary to wear latex gloves while performing confined-space rescues. Latex gloves are designed to prevent the transmission of certain diseases while you are working on a patient. If you expect to come in contact with the victim or bodily fluids from the victim, wear latex gloves under your leather gloves. The outer gloves must be discarded because of possible contamination.

Foot protection is a fairly simple matter. Steel-toed work boots with soles designed to provide traction on most surfaces are the ideal, but if not available, look for the hazards that are present and protect your feet against them. Firefighter boots will provide a level of protection from water as will waterproofed leather boots, but not against chemical contamination. This is when either preplanning or a good size-up becomes important. If you are faced with a chemical contaminant, then match the foot protection to the hazard. Rescuers may be able to wear any boots that are available, or they may have to locate safer boots or wear boot covers. The best method of decontamination is to prevent contact with the contaminants. If that is not possible and you contaminate the boots or any personal protective equipment, then you will either have to decontaminate or dispose of it. It is more effective to dispose of cheap chemical-resistive boots than expensive firefighter boots.

FIGURE 8-12

Shown here (clockwise from the top left) are a firefighter's glove, a chemically resistant glove, a rope glove and a latex glove. Each glove has a purpose and limitations to it.

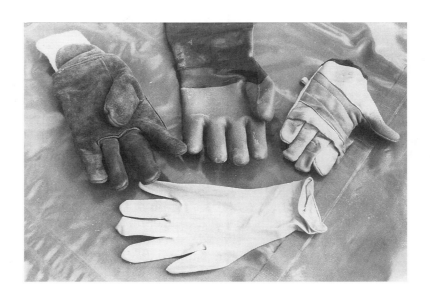

▶ **U.S. Environmental Protection Agency (EPA)**

the federal agency responsible for developing and enforcing environmental regulations.

▶ **Level A**

the highest level of skin and respiratory protective equipment for exposure to hazardous materials.

▶ **Level B**

the second highest level of skin and respiratory protective equipment for exposure to hazardous materials.

▶ **Level C**

the third level of skin and respiratory protective equipment for exposure to hazardous materials.

▶ **Level D**

the lowest level of skin and respiratory protective equipment for exposure to hazardous materials.

▶ **Permeation**

the reduction of protective properties of chemical protective clothing caused by the movement of the contaminant on a molecular level.

Body protection is also an important factor to consider when identifying hazards, and includes protecting the legs and arms against thermal (both heat and cold), chemical, or mechanical hazards. Thermal protection would require equipment that shielded the body by either insulating it or reflecting heat away from it. If thermal protection is required, this particular rescue may be highly specialized and beyond your capabilities. Chemical protection for the body would be the same as that worn at a hazardous materials incident. Varying levels of protection are available for keeping chemicals from contacting the skin and thereby avoiding either absorption of the chemical through the skin or direct skin contact, which in itself can cause damage to the skin.

The **U.S. Environmental Protection Agency (EPA)** classifies chemical protective clothing at levels A, B, C, or D. **Level A** is the highest level of both skin and respiratory protection, providing protection from hazardous chemical vapors via either skin contact with the vapors or inhalation of the vapors. Level A suits are basically gas-tight suits with either an SCBA worn underneath or an SAR and an escape bottle. It would be extremely difficult to wear both Level A protective clothing and a harness for confined-space rescue. The gas-tight suit requirements and the respiratory protection under the suit would make it close to impossible to wear a harness over the suit (you need to keep the suit gas tight), and if you could manage to get a harness to go over the suit you may very well damage it. **Level B** protection would be easier to use and is basically a splash suit with the SCBA or SAR worn outside of it. Wearing level B protection you could don the clothing, put on a harness, and then put on the SCBA or SAR. This method is easier to use, but the level of skin protection is also reduced. **Level C** and **Level D** would not be expected to be used for entry into a confined space for rescue because Level C is a splash suit with an air purifying respirator and provides only limited protection against inhalation of chemical vapors and no protection against oxygen-deficient atmospheres. Level D clothing is no more protective than ordinary street or work clothes. Mechanical protection would be required when the space has unusually sharp edges, as shown in Figure 8-13. You would need to wear garments that provide protection from puncture and cutting, usually to the arm, hand, and leg.

Chemical protective clothing is susceptible to the passage of chemicals by three different methods: **permeation, penetration,** and **degradation.** Permeation is the movement of a chemical through the protective clothing on a molecular level, as shown in Figure 8-14, which means that the chemical can move directly through the fabric of the protective clothing without leaving any visible damage. No single material can resist permeation to all chemicals; therefore, the selection of protective clothing must be made on the basis of the chemical involved and on the ability of the protective clothing to resist permeation. Penetration is the movement of the chemical through openings in the protective clothing, as shown in Figure 8-15. The penetrations in the suit may be the openings through which a person enters the suit or they may be seams and other penetrations

FIGURE 8-13

Sharp edges such as the weirs shown here require responders to protect themselves from cuts and puncture wounds.

▶ **Penetration**

the reduction of protective properties of chemical protective clothing which can occur due to zippers, seams, and other openings in the protective clothing.

▶ **Degradation**

the reduction of protective properties of chemical protective clothing by mechanical, thermal, or chemical means with a loss of integrity of the garment.

| NOTE: When you are dealing with a chemical hazard, your own level of training and experience will dictate your capabilities.

that occur when the suit is manufactured. Depending on the harmful nature of the chemical and its physical state (solid, liquid, or gas), the number and type of penetrations will affect the performance of the suit. Manufacturers may tape seams to eliminate the penetrations or provide gas-tight zippers, such as in a Level A suit, and emergency responders may minimize the effects of penetrations by duct taping zippers, cuffs, and other openings. Degradation is the failure of the suit by tearing, dissolving, or wearing away of the fabric, as shown in Figure 8-16. Exposure to a chemical may actually degrade the suit to the point that it begins to come apart, whereas tearing may occur due to stresses on the fabric or seams and the suit may be worn through by brushing against a rough or sharp surface. Regardless of how the integrity of the suit may be compromised, you must realize that no single chemical protective suit can handle all chemical hazards. Once again, preplanning a confined-space entry site can provide big dividends in terms of safety and timely operations at a rescue.

Using chemical protective clothing and being potentially exposed to a chemical hazard require the use of decontamination techniques to remove or reduce the hazard to acceptable levels. At times the best decontamination efforts will only reduce the chemical hazard to the point that people can safely remove the protective equipment and then dispose of it. When you are dealing with a chemical hazard, your own level of training and experience will dictate your capabilities. This book is not intended to provide detailed information about hazardous materials and the appropriate level of training that is required. Know what your level of training allows you to do in advance of an emergency involving hazardous materials and work within those limits.

Earlier in this chapter we discussed heat stress. Keep in mind that some protective clothing can greatly add to the potential for heat-related injuries to occur. Adding layers of required protective clothing will affect your operations to at least some extent. Keep your tactical goals and objectives realistic in relation to the limits that are presented by the use of protective equipment.

FIGURE 8-14

Illustration of permeation.

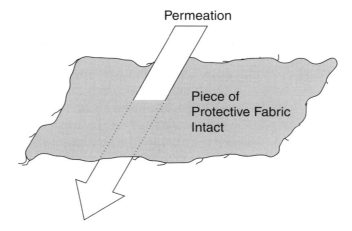

FIGURE 8-15

Illustration of penetration.

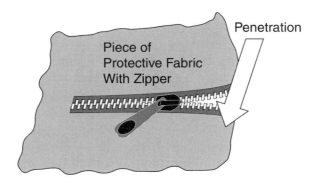

FIGURE 8-16

Illustration of degradation.

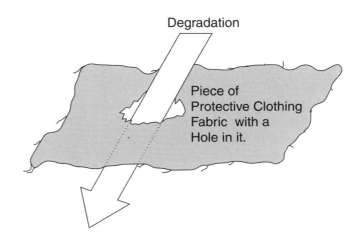

NOISE

When most people think of emergency work they do not consider noise a hazard. In most cases emergency responders consider it a part of the emergency. Confined spaces add some other considerations to noise and hearing protection. To begin, the noise within the area may be amplified by the space itself. The old expression, "an empty barrel makes the most noise," points that out. Empty containers allow for noise to echo within the space because there is nothing to absorb it. Operating within a large, bulk storage tank you may find that speaking at a normal volume becomes so loud as the sound echoes around the tank that it seems people are shouting. At other times the echo disrupts the sound of the person speaking and you cannot understand what is being said. In either case, you are creating noise that will affect your operation. Noise that interferes with communications presents an acute danger in that you may not be able to speak with entry team members or warn them of danger. If you have to change your action plan after rescuers are within the space or evacuate them quickly because of some new danger, you will be unable because the noise is too great.

The amplified noise creates both an acute danger and a chronic risk to the people exposed to it. To measure the intensity of sound, a scale called a **decibel** is used. The decibel is actually a measure of the pressure of sound and the greater the number of decibels, the greater the sound pressure and the greater the potential for damage from the sound. The pressure of the sound can be handled in several ways. Obviously, the best way is to eliminate it, but that may not be possible. If you cannot eliminate the sound then you need to reduce the level of it. This reduction can take several forms. You can use sound-absorbing materials to dampen it at the source or you can wear hearing protection to reduce the number of decibels that reach the ear. Dampening sound at the source may not be practical, in which case you will have to rely on hearing protection. Hearing protection can consist of earplugs or special earmuffs that provide the needed protection. When you are confronted with noise as a hazard that needs to be controlled at an emergency, realize that it is a problem that must be solved. Noise that interferes with communications may require the use of visual signals or specific sound signals such as using sirens or air horns for emergency evacuation alerts. Noise that will affect hearing requires that people be provided with hearing protection.

▶ **Decibel**

a unit for measuring sound intensity.

SAFETY
Hearing protection can consist of earplugs or special earmuffs that provide the needed protection.

CASE STUDY

An operator at a sewage treatment plant was inspecting filter tanks as part of his daily duties. The filter tanks are 15 feet wide by 24 feet long by 12 feet deep and contain a bed of filtering material at the bottom of the tank (anthracite coal), as shown in Figure 8-17. It is the last step in a three-stage treatment process. The operator, who is 55 years old and has over 20 years experience, is required to drain the tank, climb into the tank using a ladder, inspect the filter material, climb back out of the tank, and then put the filter tank back in operation. Of the twelve filter tanks at the plant, the operator had completed inspecting four. This inspection was a daily practice and the operator had inspected the tanks two days prior to this.

You are called to the treatment plant just after 11:00 A.M. and upon arrival are told by workers at the plant that the operator is lying, on his back, at the bottom of a filter tank. He is visible from the top of

FIGURE 8-17

FIGURE 8-17

Diagram of the case study. This space was below grade, and the victim was found lying on top of the filtering medium.

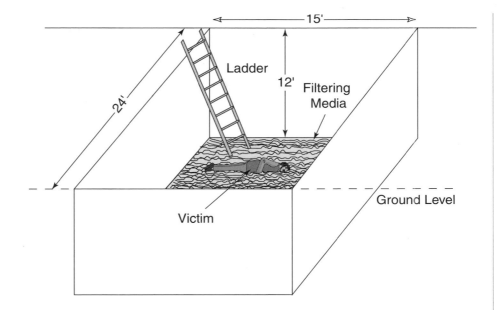

the tank, has a bleeding wound on the side of his head, and is unresponsive. The weather conditions are clear, no wind, high humidity, and the temperature is already 80°F and is expected to top 90°F.

Questions to Consider

1. Is there a potential for heat-related injuries to occur at this incident?
2. What type of respiratory protective equipment would you recommend?
3. Do the rescuers need to wear a rescue harness?
4. Are there any other safety concerns that you can readily identify?

Answers

1. With both the temperature and the humidity high, personnel will begin to feel the effects of heat. If the rescuers are required to wear impervious clothing, the effects will occur faster, which means that you must anticipate how long the operation is going to take and be prepared to provide relief for the rescuers. Additionally, this tank is open on top and is exposed to the sun. The filter material is black coal and will heat up quickly under the sun's rays. As personnel are setting up equipment for the rescue operation, have the rescue team begin hydrating. You can do this while you brief the team members on how they are to rescue the victim. Minimize the amount of protective clothing that the rescuers must wear. If you have coveralls available, consider if they will be effective, otherwise chemical protective clothing such as Tyvek™ suits or firefighters' protective clothing can be utilized. Helmets, boots, and gloves are a must in this circumstance, but any protective clothing will add to the heat load on the rescuer and must be considered.

2. You have two choices for respiratory protective equipment—SCBA or SAR. What choice will you make? An SCBA will be effective in this case because the opening is not limited and the SCBA can be used quicker than the SAR. However, you must look at the emergency as it is presented to you. If special circumstances require an SAR in this case and if that is what you recognized as a problem, then handle the problem.

3. The rescuers are going below grade and you must be sure that you can rescue them as quickly as possible. Having the rescuers wear a harness and attaching it to a lifting device means that you will be able to get your personnel out of the tank quickly without having to send more people into the tank.

4. Other safety concerns that you should have at this incident include the stability of the filtering media, the use of ladders, the biohazard precautions, and the possibility that some hazardous chemical is present. The filtering medium is stable, does not contain voids, and is piled on the bottom of the tank—this information you discovered during preplanning. A ladder is already leading into the space, but how good is it? Can it support the weight of your rescuers, their equipment, and the victim? It may be better to use a portable, fire service ladder which is designed to handle heavy loads. Biohazard precautions need to be addressed here because the victim is bleeding and you are in a sewage treatment plant. You may easily recognize the hazard related to treating a victim, but what precautions are required because of the presence or potential presence of the sewage? Depending on the stage of treatment, this tank may be contaminated and you will have to protect your rescuers against that hazard. Hazardous chemicals may be a possibility when trying to determine what has caused this accident, but is it a probability? At this stage of treatment, the water has passed through a lot of other checks and processes. If this water was just coming into the plant, the presence of hazardous chemicals would be a greater possibility, especially with other indications such as positive meter readings or chemical odors. You still question the people at the scene about possible actions that might have introduced a hazard. Your main responsibility is to rescue the victim, and part of that rescue process requires exposure to risk and management of the exposure.

In this particular case, the victim suffered a heart attack from chronic cardiac disease and emphysema. The victim happened to be climbing the ladder when he collapsed and fell back, striking his head. Because this incident was unwitnessed, as the victim was working alone, you had little information on which to base your decisions. Wearing respiratory protection and other protective equipment did not significantly add time to the rescue, but it did increase incident safety and reduce risk. Nothing would have been gained by not wearing the equipment.

■ SUMMARY

The safety of personnel at a confined-space rescue operation must be of paramount importance to every member of the rescue team. The incident commander is seen by many as the person who has the primary responsibility for safety during the incident, but everyone shares that responsibility to some degree. Addressing safety issues for emergency response begins long before the actual response. The selection, use, and training with equipment is an integral part of safety during an emergency. Identifying and remedying issues such as hypothermia, heat related injuries, or personal protective equipment requires preparation both before an emergency and adherence to those preparations during an emergency. Safety must be ingrained into the emergency organization and its operations. Without an organizational culture that considers safety to be an indispensable part of that organization, safety will be an afterthought. With safety as part of your culture, you can identify the hazards and either eliminate them or, when that is not possible, protect people from the hazard. From beginning to end, your emergency operation should consider safety as one of the most critical strategic factors.

■ REVIEW QUESTIONS

1. Describe the conditions (temperature, humidity, etc.) under which heat stress injuries can occur.
2. Describe the conditions under which hypothermia can occur.
3. Identify how temperature-related injuries can be prevented and the effects those prevention measures can have on your rescue operations.
4. NFPA Standard 1983 covers life safety rope and related equipment. Describe the value of using equipment which has been designed and built to nationally recognized standards (ANSI, NFPA, OSHA).
5. Describe the purpose and limitations of at least three different pieces of protective clothing.
6. The movement of a chemical through protective clothing on a molecular level is ____.
   a. Penetration
   b. Degradation
   c. Permeation
   d. None of the above
7. Hearing protection is not needed at a confined-space emergency. True or False?
8. It is an acceptable practice to enter a confined space by removing your SCBA and then placing it back on after you have entered. True or False?

9. Using the windchill chart in Figure 8-7, if the temperature was 40°F and the wind speed was 30 mph, the resulting temperature effect would be _____°F.

    a. 37
    b. 22
    c. 13
    d. 0

10. When the potential for heat stress injuries is present, the entry team should be required to drink 12 to 16 ounces of water or sports drink only before entry. True or False?

CHAPTER

9 Rescue

OBJECTIVES

After completing this chapter, the reader should be able to:

- list the nine-step process for confined-space rescue.
- categorize the different types of equipment used for confined-space rescue and the limitations and standards which apply.
- describe initial scene operations.
- recognize the basic considerations you must have for victim assessment, victim stabilization, and victim removal.
- categorize operations as defensive or offensive.

RESCUE CONSIDERATIONS

You are a trained rescue technician and you have been called to a confined-space emergency. As you arrive at the scene many considerations you have been trained to recognize and act upon flood through your mind. Should you begin to ventilate or should you use a meter in the space? The reason you are involved in emergency services is to help people and now you will have that opportunity, but first you need to start at some logical point. Confined-space rescue is basically a nine-step process, as follows:

1. establish command and take control of the incident
2. identify the type of rescue problem
3. perform hazard and risk assessment
4. identify rescue objectives
5. identify resource needs
6. develop an action plan
7. implement the action plan
8. evaluate the effectiveness of the action plan
9. terminate the incident

Establish Command and Take Control of the Incident

It will be difficult if not impossible to do anything to help the victim(s) of an emergency if no one is in charge of the scene. Without a person in command, as shown in Figure 9-1, no other part of the nine-step process can take place effectively. Picture a scene with everyone shouting orders and no one following them. What would you expect to get accomplished? Chances are that at such a scene the only result would be an increase in noise volume from the voices of all the people giving orders. You came to this emergency to make a difference. You came to help people and now it has fallen apart. What a frustrating situation this could be, unless that first trained emergency responder at the scene establishes command and takes control of the incident. Now, with someone in charge, orders are coordinated, actions are based on the outcome they will have on other actions already taking place, and the best care will be provided for the victim. Handling an emergency is not an easy task, but as an emergency responder this is your job and you are expected to act as a professional. How much faith would you have in a doctor who works in an emergency room if you arrived with a badly cut hand and several doctors began to shout out conflicting orders to create confusion there? Chances are that you would take your business elsewhere.

Identify the Type of Rescue Problem

If you get called to an emergency that involves the rescue of an injured or trapped person, will it always be considered a confined-space accident? No, but you may have still been dispatched to a

FIGURE 9-1

Effective control of the rescue begins with establishing command. Failing to establish command will affect all additional operations.

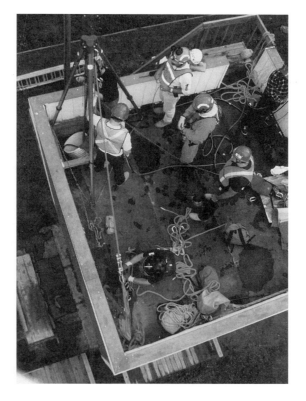

FIGURE 9-2

Identify the rescue problem. Is the victim in a storm sewer, an above-ground tank, or some other type of confined space?

confined-space incident. On receipt of the alarm, dispatchers do the best job they can with the information they are given. Based on the initial information you were given, begin a mental size-up of the emergency. When you get to the location, look further at the problem and determine the type of rescue needed, as shown in Figure 9-2, and your capabilities of performing the needed tasks. You may have been dispatched to a reported confined-space rescue only to find out instead that it is a high-angle rescue or a hazardous materials incident which requires rescue and there is no confined space involved. Last of all make sure that you are required to make a rescue. By the time you arrive at the scene the victim may be out of the confined space and not require rescue.

Hazard and risk assessment begins upon dispatch to the incident and is an ongoing process during the incident. Here the rescue team is reviewing the confined-space entry permit.

Perform a Hazard and Risk Assessment

In performing a hazard and risk assessment, you will be continuing your size-up and looking for the strategic factors that will affect your operations. Reviewing on-site confined-space permits (as shown in Figure 9-3), questioning people who witnessed the incident or who were present when it was discovered, and using monitoring equipment are just a few of the means you will use to identify the hazards present. This stage of the emergency is the information gathering and evaluation stage. The presence of hazardous materials within the space, an oxygen-deficient atmosphere, or any other significant hazard must be recognized; but you must also judge how those hazards will affect your incident. Your decisions will be based on what you determine to be the circumstances of the emergency. If the hazard can be easily corrected or controlled, the risk is reduced; but if it cannot be controlled, you must manage those risks in other ways.

Identify Rescue Objectives

Knowing that an oxygen-deficient atmosphere is present, you would be required to wear an SCBA, ventilate the space, and continuously monitor the atmosphere. This fact presents you with three pieces of information and several decisions to make. The hazard and risk assessment you should be performing is going to lead you in a particular direction in terms of critical factors. Based on those critical factors you must begin identifying what you can do to rescue the victim, as shown in Figure 9-4, and protect other people at the scene. Keep in mind the order of importance of incident priorities: life safety, incident stabilization, and then property conservation.

Clearly your goal is to rescue the victim and make sure that no one is injured during that time, but goals are only general statements of what you want to achieve. Objectives are the specific and measurable means you will use to accomplish your goals and you can carry out only so many objectives. As you gather information, you

FIGURE 9-4

Identify your rescue objects. Is it necessary to enter the space to perform the rescue?

will begin to realize that some objectives will have more impact than others. Sort out the different ones by identifying them as either easy or difficult to achieve. Then sort them further, as having high impact or low impact on the outcome of the emergency. Those items that you have identified as easy to achieve and having high impact should receive immediate attention. Upon establishing the plan of action, you will have set your objectives for the emergency. One word of caution is in order, however. Simply because you have identified something as easy to achieve and having high impact does not mean that it will work. Be prepared to review your objectives occasionally to determine if they are working as planned.

Identify Resource Needs

With your objectives set, you will need to determine who is going to be assigned to carry out the tasks that need to be accomplished and what equipment they will need, as shown in Figure 9-5. People and the equipment being used are considered resources. The status of your resources falls into three categories: available for use, in use, and not available. Beyond the status of your resources are other limits on resources that you need to recognize. If what you are trying to accomplish exceeds those limits, you are setting your operation up for failure. That failure can result in wasted time, the loss of equipment, and even death for the victim and rescuers. The more critical the failure, the more serious the consequences, so the selection of which resources go to what tasks gains importance based on the impact that a failure will have on the operation. At times that failure will be a result of equipment being used beyond its design limits; at other times it may be a result of failing to provide proper maintenance, and at other times it will be the result of people who do not know how to use a particular piece of equipment. Emergency responders who are tired, who are not trained for the tasks they are expected to perform, or who otherwise are ill-prepared for the work they will be doing will be a tactical disadvantage. When resources

FIGURE 9-5

FIGURE 9-5

What resources do you need, and what resources do you have available?

are inadequate call for those you need. When they are not available, accept the circumstance as a critical factor and reflect it in your action plan.

Develop an Action Plan

An old axiom states, "When you fail to plan, you plan to fail." An emergency operation requires an interdependency on many different parts of the operation working together, so a plan is an absolute necessity. The plan can be simple in that the requirements of your operation are simple or it can be complex based on the jobs to be performed. Regardless of the complexity, all emergency operations require a plan. At times the plan may involve a series of tasks in sequence, as shown in Figure 9-6. Other situations may involve simultaneous tasks to achieve the objectives. Without a plan, however, the coordination and control needed to ensure success will be lost.

Implement the Action Plan

At this point you know what you want to do, you know what you need to succeed, and you have a plan. Now you may implement your plan, as shown in Figure 9-7. Be sure that everyone who is affected by the plan knows their specific role and when they are expected to perform it.

Evaluate the Effectiveness of the Action Plan

What you expect to accomplish with your plan and what you actually accomplish can be two different things. It is rare that your action plan will proceed exactly as you set it out. If you do not take the

FIGURE 9-6

In developing an action plan, identify the sequence of actions needed to complete your operation. Equipment such as retrieval equipment may need to be in place prior to entry.

FIGURE 9-7

Implement the action. Once you are set up and ready to go, begin your operation.

time to review the effectiveness of your plan and make adjustments as needed, you will not complete the objectives you set out to achieve. This evaluation of your action plan is an ongoing process, as shown in Figure 9-8, and it is not completed until the incident is terminated.

FIGURE 9-8

Evaluate the effectiveness of your operation. How are your personnel, equipment, and other resources performing?

FIGURE 9-9

Debriefing rescue workers after an incident can provide valuable information for future use. These students are being debriefed after completing a training exercise.

Terminate the Incident

The emergency is not over until you have provided closure to the situation. This closure includes debriefing your personnel to determine the factual circumstances of the incident (as shown in Figure 9-9), letting emergency responders know of any special circumstances that may be important (e.g., signs and symptoms of exposure to hazardous materials present at the emergency, condition of victims, critical incident stress considerations), and providing a follow-up incident analysis and critique to determine the effectiveness of the operation and any needed improvements for future incidents. Termination activities may also include maintenance of equipment, record keeping, and follow-up enforcement activities with other agencies for prevention of similar incidents.

One particularly important area that should be included in debriefing is to review the need to address critical incident stress. The people who worked at the emergency are just that, people. The rescuers also have emotions, and the incident may evoke many different ones which are normal reactions to abnormal situations. The death of the victim, an emergency worker being seriously hurt, or a prolonged operation with little chance of success are some examples of the type of situation that can lead to critical incident stress. Recognition and early intervention can prevent long-term effects from such stress. If you have critical incident stress teams available in your area, you should know how to activate them and what can be done for the benefit of your people.

EQUIPMENT

The type of equipment used for confined-space rescue will vary from the simple to the sophisticated. Preplanning will help to determine your particular needs, but it is important to keep in mind that the simpler the equipment is to set up and use, the simpler it can be to make a rescue. The simplest equipment will not always be right for the job, but it can be used by many people without large-scale investments in training. It is the equivalent of the difference between using a ground ladder to rescue a person from a window and using an aerial ladder. Under the right conditions, both ladders will do the job, but the initial training requirements for the ground ladder are much less demanding than the aerial ladder. In addition, there is a need to provide periodic training to maintain a satisfactory level of proficiency, so look at the tools you will need to do the job, look at how you will have to train to be proficient, and look at the people you have available. When you can match those three items, you will be able to use the equipment properly.

Previous chapters discussed personal protective equipment, harnesses, blowers, and monitoring equipment. We will not repeat that information here except to remind you that you must have that equipment available. What needs to be addressed here is the other equipment you may already have or will need to acquire.

Tripods are a good example of equipment that is unique to confined-space rescue, as shown in Figure 9-10. Tripods come in different sizes based in part on the height, footprint, and lifting capacity of the unit. There are important distinctions between different tripods, including the number of anchor points on the head of the tripod, the ability to attach lifting devices to the legs of the unit, the ability to adjust the height of it, and the lifting capacity.

Tripods are meant for use by people. Lowering or raising equipment or other loads into and out of a confined space with a tripod should be forbidden. During an emergency you will have to depend on every part of the tripod working to its capacity. Using the tripod for loads other than people (especially heavy loads) can lead to its failure. When a load is applied to the tripod, it must be transmitted to the ground either through the axis that runs through the center of the tripod or through the axis that runs down the center of each leg. In other words, if you attempt to use a tripod to retrieve a victim and the direction of the pull is in any direction but along the axis of either the tripod or each leg, you will pull the tripod over. To keep the

---

NOTE: The type of equipment used for confined-space rescue will vary from the simple to the sophisticated.

▶ **Tripod**

three-legged retrieval support devices.

SAFETY
Tripods are meant for use by people. Lowering or raising equipment or other loads into and out of a confined space with a tripod should be forbidden.

FIGURE 9-10

Among the various pieces of equipment that can be used for confined-space rescue is a tripod. Know the capabilities and limits of your equipment.

load applied in the proper direction, you must use either retrieval equipment specifically designed for use with the tripod or anchor points and change-of-direction pulleys which keep the load aligned with the tripod until it is transferred to the change-of-direction pulley. This process is simpler than it sounds, but you must be aware of how you load your retrieval equipment.

In the event that it were possible to safely use a ladder to enter a confined space, you would still want to use a tripod and attach a retrieval line on any rescuers entering the confined space. A retrieval line would allow you to rescue the person wearing it, without entering the space. The retrieval equipment used with a tripod may also be designed to limit a free fall of a person climbing up or down an access ladder in the confined space. You would also want at least a retrieval line for each person entering and a safety line while they were being raised or lowered, which would require a minimum of three lines (one for each rescuer after they are in the space and a safety line for lifting) and three attachment points at the head of the tripod. Once two rescuers were in the confined space, there would be two retrieval lines in use and a spare third line. When the victim was ready to be removed, you could use the spare line and the retrieval line to take one of the rescuers out of the space and then send both lines back into the space to be used to recover the victim. After the victim was removed, you would send one line back into the space as a safety line to be used by the rescuer left there. Using a safety line seems redundant, but it is only a simple matter of ensuring that if anything happens to one line or to the retrieval equipment on it, you can recover your personnel and the victim safely.

SAFETY

Using a safety line seems redundant, but it is only a simple matter of ensuring that if anything happens to one line or to the retrieval equipment on it, you can recover your personnel and the victim safely.

FIGURE 9-11

An aerial ladder raised in anticipation of use for a confined-space rescue. Make sure that you know the limitations of your equipment.

When specifying tripods for purchase, look at the length of the cable on any retrieval system, the maximum workload, its weight, and any additional features such as the ability to adjust the leg height and chains or cables to keep the legs from spreading. If you have some idea what rescue situations you will be facing, you can purchase equipment that will fit your needs.

If you do not have a tripod or cannot use a tripod to raise and lower people for rescue purposes, other equipment may be of use. It is possible to use ground ladders to build A-frames and aerial devices (ladders, tower ladders, Squrts™, etc.) to provide gin poles for lifting, as shown in Figure 9-11. Before you consider doing that, however, you must absolutely know the design limits of the equipment you will be using. Ground ladders that are used to make an A-frame must be able to carry the accompanying load. In addition, the manner in which you fasten the ladders will affect the capacity of the system you are building. What good are ladders that can support several hundred pounds if you use rope for lashing that can only take 50 pounds of weight?

Aerial ladders have different loads that they are capable of carrying. Newer aerial ladders may support fairly large loads at full extension of the ladder and at low elevations, but what can your ladder support? Besides the load you will be applying to an aerial ladder, remember that it is a mechanically powered piece of equipment. Mechanically powered equipment poses the additional danger of injuring people because of the forces it can apply in the event that a person becomes trapped or wedged while being lifted. The use of improvised equipment should only be considered after you know the limits of the system you want to create, which means that you must preplan and

SAFETY
Mechanically powered equipment poses the additional danger of injuring people because of the forces it can apply in the event that a person becomes trapped or wedged while being lifted.

FIGURE 9-12

Know what your ropes are designed for and how they were meant to be used.

► **Laid ropes**

a type of rope that is made by twisting smaller strands of rope together.

► **Braided ropes**

a type of rope constructed by interweaving the strands of the rope together.

► **Kernmantle ropes**

a type of rope made with a load-bearing core (kern) and an outer sheath or braided cover (mantle).

train using the equipment and methods that you want to employ. Creating new and untested devices in the field is dangerous and should be discouraged. If you need specific information about the ability of aerial equipment, ground ladders, and other equipment, contact the manufacturer and get the design specifications for the equipment.

A variety of ropes can be used at confined-space rescues, as shown in Figure 9-12. This book is not intended to teach rope rescue techniques and will only briefly discuss ropes and associated equipment. It is necessary for rescuers to understand some basic rope terms and the types of rope that are available. To begin, ropes are made by varying methods, including **laid ropes, braided ropes,** and **kernmantle ropes.** Laid ropes are made by twisting the strands of rope together. All strands of the laid rope contribute equally to its strength. Braided ropes are simply strands that are braided together and made either with or without an outer jacket. Kernmantle ropes use an inner core called a kern and a protective outer jacket called a mantle. The kern carries most the load placed on the rope, up to 75 percent of the load, and is protected by the mantle.

In addition to its construction, rope may be either static or dynamic. Static ropes have little stretch and are most often used for rescue, whereas dynamic ropes are designed to stretch and are typically used as shock-absorbing protection such as from a fall. Ropes will be selected as either life safety or utility rope. Ropes used for life safety must be fabricated so that there is a continuous filament fiber within it to ensure that it will not separate under its working load. Utility ropes are not designed for life safety use and are only intended to haul equipment, to serve as tag lines, or to carry loads other than people. Life safety rope is intended to carry the weight of the rescuers and any victims and must be made with a high degree of reliability and safety in mind. NFPA Standard 1983 defines life safety rope as, "Rope dedicated solely for the purpose of constructing lines for supporting people during rescue, fire fighting or other emergency operations, or during training evolutions." Do not allow life safety ropes and utility ropes to be intermingled unless it is clearly obvious which

FIGURE 9-13

Using this rope and harness, could you retrieve this rescuer if he became injured or entangled?

FIGURE 9-14

A carabiner used to connect a figure 8 descender to a sling.

rope is for which purpose, and never use utility rope for life safety purposes. You must also consider the need to secure all people with two ropes. One line would function as the main line or working line, as shown in Figure 9-13, while the second rope would be a safety line in case of failure or damage to the main line or any of its hardware.

In addition to ropes, particular rope hardware devices are used with the life safety rope. This hardware must also meet NFPA Standard 1983 and consist of ascent devices, descent devices, pulleys, carabiners, and other equipment used to protect, connect, and control them, as shown in Figure 9-14. These devices can be used to at-

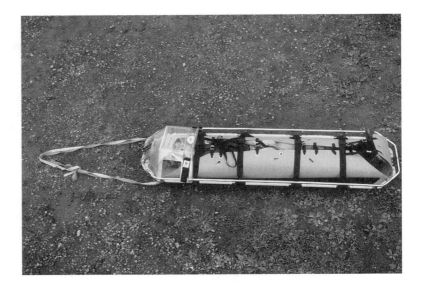

FIGURE 9-15

A Sked™ stretcher placed in a stokes basket.

FIGURE 9-16

Ropes, slings, pulleys, and other rope equipment for use during a confined-space rescue.

tach the rope to slings for anchor points, to other rescue equipment such as a Sked™ (as shown in Figure 9-15), to harnesses (as shown in Figure 9-16), and to construct systems that provide mechanical advantage when moving a load. Rope hardware can also be used to control descent as the rope is fed through a descending device, to protect the rope from chaffing, or to change the direction of the pull on the rope so that it can pass around an obstruction or be pulled more efficiently.

The use of rope hardware to provide mechanical advantage can range from a simple change-of-direction pulley to allow more people to pull on the rope, as shown in Figure 9-17, to the construction

FIGURE 9-17

FIGURE 9-17

A pulley being used to change the direction of pull on the rope by 90°.

NOTE: Utility ropes are not meant to be used for life safety and therefore only need to be strong enough to haul any equipment that may need to be raised or lowered.

of two to one or three to one systems. Using rope for confined-rescue purposes can be invaluable, but the more complex the systems you need to build, the more demanding the training. Regardless of your intended use for rope during confined-space rescue, first train to the level of proficiency that is required and then continue to train to maintain your proficiency. Training to this level and then maintaining it is not an easy task. In fact, if you plan to use rope for confined-space rescue then you need to clearly set out what you expect to accomplish before you purchase equipment and begin training.

Utility ropes are not meant to be used for life safety and therefore only need to be strong enough to haul any equipment that may need to be raised or lowered. Life safety ropes should not be used as utility ropes because equipment may have sharp edges, rough spots, or contaminants that can damage the rope. Utility ropes should be constructed of materials that allow you to tie knots easily and that will hold the knot.

Equipment that you will need for confined-space rescue work, as shown in Figures 9-18 and 9-19, includes rope yokes which are designed for use with either Class III harnesses with shoulder-mounted D-rings or with wristlets, a rescue stretcher such as a Sked™ stretcher, backboards for immobilization, low voltage lights or other lighting equipment that is listed as safe for the environment in which it will be used or that will not become an ignition source, and communications equipment. Wristlets are designed to go around either the rescuer's or the victim's wrist and have a retrieval line attached at the center. The person is then raised or lowered by his arms extended over his head (not comfortable, but effective). Rescue stretchers specifically designed for use in technical rescue are a must, because the stretcher must fit through the opening of the space with the victim in it. Pre-formed stretchers such as Stokes baskets cannot be made smaller and are not as effective as rescue

FIGURE 9-18

Equipment for confined-space rescue laid out in a staging area in order to organize and account for the equipment.

FIGURE 9-19

A rescuer wearing a supplied air respirator with communications equipment attached to the airline and built into the facepiece.

stretchers designed to fit the victim. Backboards are an obvious requirement for immobilizing victims, but you should check your backboards to see if they will work effectively with your rescue stretchers. Electrical equipment used for lighting can be an ignition hazard. If this ignition hazard cannot be removed, the flammable atmosphere must be eliminated. You certainly do not want to electrocute your personnel or the victim. Ground fault circuit interrupters can help to prevent electrocution, but a simpler method may be to use low voltage lights.

Communications equipment must be used to keep in contact with the rescue team, but using it when wearing protective equipment may be difficult. Newer respiratory protective equipment has a built-in microphone in the facepiece, but that does not solve the problem. In order to talk, you may need to press a button or switch. If the protective clothing makes that step difficult, again you cannot communicate. At times it may be possible to maintain visual communications with the rescue team (rescue entrants) and rescue attendant or incident commander outside the space. In such a case you could use verbal messages from outside which can be answered by visual signals from the rescue entrants. Unfortunately, visual communications will not always be acceptable. Preplanning your communications needs, acquiring the equipment, and then practicing with it will teach you about the effectiveness of the solution. At times communications may become a strategic factor and require that you call additional resources to solve the problem.

Equipment is only as good as the people using it. You may have the best equipment available, but if your people do not know how to use it properly, it is useless. As this book is being written, the NFPA is developing **Standard 1670—Standard on Operations and Training for Technical Rescue.** This standard will be dividing technical rescue into different areas such as confined-space rescue, vehicle and machinery rescue, and rope rescue. The standard will also define different levels of training and competency for rescuers such as an awareness level, an operations level, and a technician level. This standard will begin to identify requirements for training that must be met to obtain the proper level of proficiency to perform different types of rescue.

Hopefully, the proposed NFPA standard will provide a basis for identifying the minimum training needs. You should think of training with your equipment as maintenance. Like maintaining the proper oil level in your car's engine, active training for confined-space rescue is a necessity. To use your equipment, you must maintain the proper level of proficiency; training is part of that maintenance. Actual emergencies provide experience that would be difficult to re-create in training; however, you can only learn from actual emergencies if you critique the operation after it is over. You must be willing to look at how your team was able to perform, what they did both right and wrong, and what could have been done better. Did your equipment perform as expected or was there a problem with the equipment or your use of it? Termination activities, as mentioned, include critiquing the incident. Documenting the incident is

NOTE: You may have the best equipment available, but if your people do not know how to use it properly, it is useless.

▶ Standard 1670—Standard on Operations and Training for Technical Rescue

a proposed NFPA standard currently being developed to define and categorize technical rescue qualifications.

part of the termination process and not only provides you a record, but also can verify information that is useful for training. If you have no incidents with personal experience from which to draw, seek out information from others who have had incidents. Look at what those other people were able to accomplish and what limitations they encountered. Use their successes or failures as a training tool.

INITIAL SCENE OPERATIONS

When you first arrive at the scene of a confined-space emergency, you will look to start operations as soon as possible. Go back to the nine steps outlined earlier in this chapter and begin to work on each step. Some may be completed sequentially, whereas others can be performed simultaneously. When you are aware of the problem, set out your objectives and start with those that are easy to achieve and have the greatest impact on the situation. Identify how many victims you actually have—people may have entered the space in a misguided attempt at rescue. Look to start ventilation and atmospheric monitoring as soon as possible. Both ventilation and monitoring may be started from outside the space without requiring any personal protective equipment. You know that your goal is to rescue the victim, but it is possible that the victim may be dead. What degree of probability is there that the victim is dead? If the victim is inside of a small manhole where high-pressure steam lines run and the space has been filled with steam at over 300°F for the past 10 or 15 minutes, then it is not only possible that the victim is dead, there is also a high degree of probability that he is. You are not performing a rescue in this case, simply a body recovery, and your objectives should reflect that fact. If you are unsure of the victim's condition, however, then your objective is going to be rescue.

Hazardous materials emergencies involve two basic types of operations: **defensive** and **offensive.** The same can be applied to confined-space emergencies. Defensive operations can be performed without entering the hazardous area and may not require personal protective equipment, as shown in Figure 9-20. Defensive operations can buy time to allow offensive operations to be put into place, such as ventilating and monitoring the space. At other times defensive operations may be all that is needed, such as removing a victim using the retrieval equipment from outside the space. Offensive operations require more equipment and resources and take longer to accomplish, as shown in Figure 9-21. You will need to have all necessary resources in place before starting an operation. Try telling one of your rescue entrants that you can lower them into the space, but until a retrieval device gets there you cannot get them out. Certainly the rescuer would invite you to lead the way into the space. Gathering your resources and setting them up takes time, which is a commodity the victim may not have, so try and buy time by getting those high-impact, easy-to-do actions started. Remember the incident priorities: life safety, incident stabilization, and property conservation. If the rescuers cannot enter the space at this time, and you can buy time for the victim to survive, you have not only taken care of life safety, but you may have stabilized the incident.

▶ **Defensive actions**

actions taken in which there is not intentional contact with the hazards of a confined space.

▶ **Offensive actons**

actions taken in which there is expected to be intentional contact with the hazards of a confined space.

FIGURE 9-20

Both defensive and offensive operations can be taken during a confined-space rescue. This chart characterizes defensive operations.

Defensive Operations

Performed without entering space or prior to entering space

Examples:

Establishing ventilation
Use of monitoring instruments while outside the space
Non-entry rescue or retrieval of victim

Advantages:

Limits risk to rescuers
May assist in stabilizing the incident prior to entry
May be the fastest and most successful way of rescuing victim
Level of knowledge and skill possessed by rescuers at lower levels

Disadvantages:

May not be effective means of rescue of victim

FIGURE 9-21

Offensive operations place greater demands on rescuers and equipment.

Offensive Operations

Performed by entering the confined space for rescue

Examples:

Use of monitoring instruments while inside the confined space
Packaging and removing victim from within space
Assisting another rescuer within the confined space

Advantages:

May be the only way to rescue the victim from the space

Disadvantages:

Requires higher level of risk to be managed
Requires higher levels of training and skill for rescuers
Requires additional equipment for entry, retrieval, and personal protection

---

SAFETY
At least one member of the confined-space team (hopefully all members) must be trained in CPR and basic first aid.

ACCESSING THE VICTIM

At least one member of the confined-space team (hopefully all members) must be trained in CPR and basic first aid. The most simple item in a primary survey for any victim is airway, breathing, and circulation (ABC). If the victim cannot maintain an airway, he cannot get air. A victim laying face down in a puddle of water at the bottom of the confined space is not maintaining an airway. No airway for a long enough period of time means no life. If the victim is maintaining an airway or you can reestablish an airway, then he may be breathing, but what can you do for a person who is not breathing?

Provide ventilations? With respiratory protection on or in a contaminated atmosphere?

Get a rescuer into the confined space, as soon as possible to begin ventilating the victim, but remember that you must either change the atmosphere or provide a self-contained source of oxygen for the victim. Removing the victim quickly may be the only answer, as shown in Figure 9-22. Circulation not only consists of pumping the blood around the body, but it also requires having blood to pump. Performing CPR in a confined space may be difficult if not impossible. In that case the victim must be removed from the space as soon as possible, which presents the issue of spinal injury. If the victim is dead then the other injuries will not be treated. If the victim can be kept alive by being removed from the space, the other injuries that may be permanent will be treated as needed. If the victim's ABCs are good, begin to perform a secondary survey. Now the injuries take precedent and you will have to prepare your patient for transportation.

VICTIM STABILIZATION

Stabilizing your victim should follow standard protocols for a trauma patient in most cases. Your victim may have fallen or may be unconscious. Well-intentioned people outside the space may tell you that the victim had a heart condition or other medical problem, but they may not be telling you that the victim fell or why he fell. The victim may indeed have collapsed from a heart problem, but he could have collapsed just as easily from a hazardous atmosphere. You must work from facts and the facts that caused the person to fall are unknown to you. You do not know how far the victim fell but you do know that he fell, and that implies a potential trauma patient. The use of spinal immobilization is indicated, including a backboard.

FIGURE 9-22

How rapidly your victim must be removed from the confined space should be one of the primary considerations in packaging your victim.

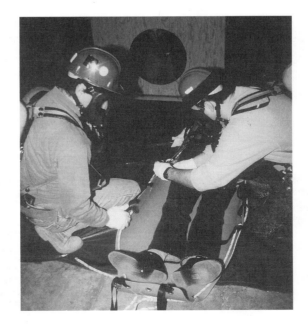

Your rescue stretcher should permit the use of a backboard. You must train with and know this equipment.

VICTIM REMOVAL

Remove the victim from the confined space *in a manner that is best for the victim*—the process of removing the victim should not be more hazardous than letting the victim stay in the space while you prepare a safer way out. Taking the victim out sideways, as shown in Figure 9-23, is generally better than lifting him out vertically, but if the vertical method is the only possibility, make sure you have the proper equipment to do the job. The position of the rescue stretcher is also important. If the stretcher can be kept horizontal, even if it must be lifted, the victim will be much more comfortable, will stay in place on both the backboard and the stretcher, and does not face the risk of falling out of the stretcher (barring a failure of the equipment). When placing the stretcher vertically, as shown in Figures 9-24 and 9-25, make sure that the victim is properly positioned and secured to the stretcher. Ensure that the stretcher is designed to be used in the vertical position and that you have fastened the victim to the stretcher as required for a vertical lift. Prior training will pay off here, as the degree of competency needed to package and lift a patient must be 100 percent. It is always a good idea to check any stretcher that must be lifted. Before you begin to raise the stretcher out of the space, lift it about a foot off the ground and allow the full weight of the victim and stretcher to load the equipment. Check all fasteners, knots, straps, and so forth to be sure they are holding and the victim is not slipping out or in danger of slipping out. Only if everything checks out okay should the lift begin. If it does not appear to be satisfactory, bring the stretcher back down and fix the problem. Again, prior training will be of help because you will not

FIGURE  9-23

The location of victims and your access to them will affect how you can remove them.

FIGURE 9-24

The narrow width of these stairs, combined with obstructions such as this pipe flange, may make it impossible to carry a victim to the ground.

FIGURE 9-25

The location of this manhole, on a steep hillside, combined with the narrow opening will require you to remove the victim in a vertical position.

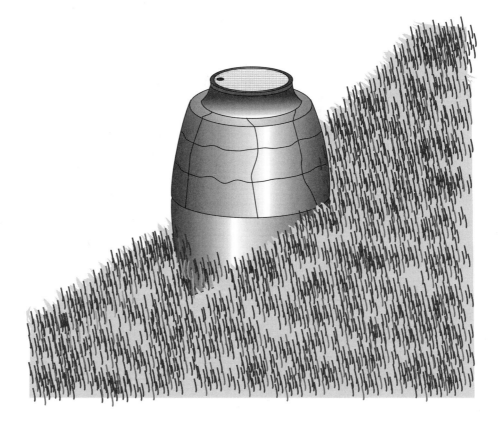

have a checklist in the space and you will therefore have to rely on what you learned in training.

■ SUMMARY

You have been called to a confined-space emergency because you are trained, prepared, and equipped to deal with the problem; so you are expected to act in a calm, professional manner. Knowing, in advance, at least a basic concept of what needs to be done will go a long way in getting your operation started. Begin by establishing command and taking control of the emergency. Identify the problem and the associated hazards. Determine the objectives, the resources you will need, and the necessary actions to achieve those objectives. Begin the operation using your plan and continuously evaluate your plan's effectiveness. When you have completed the operations, terminate the incident by gathering information about it, recording that information, and then honestly looking to see what lessons you can learn. This process all sounds simple enough, but you must also know, in advance, what limits your resources can place on you. Preplanning, training, and maintaining a standard of reliability (for both people and equipment) will develop a reliable level of competence.

■ REVIEW QUESTIONS

1. Identify and briefly explain the nine steps for confined-space rescue.

2. Explain how tripods should be loaded for lowering or raising a rescuer or victim. If the load cannot be applied in the proper direction, explain how it can be transferred so that it is applied in the proper direction. You may use a diagram to explain, but you must show the direction in which the load is being applied.

3. What NFPA standard addresses life safety rope?

4. Explain the difference between defensive and offensive operations and the value of each type at a confined-space emergency.

5. Tripods for confined-space rescue can be used to raise and lower both people and for equipment hoisting. True or False?

6. OSHA Standard 1910.146 requires how many rescue team members to be trained in CPR and first aid?

   a. 1
   b. 2
   c. 3
   d. More than 4

7. The best rescue equipment available is only as good as the people trained to use it. True or False?

10 Standard Operating Procedures

OBJECTIVES

After completing this chapter, the reader should be able to:

- identify the purpose of standard operating procedures.
- identify the value of checklists.
- identify the limitations of standard operating procedures.

DEVELOPING STANDARD OPERATING PROCEDURES

▶ **Standard operating proce-**
dures (SOPs)

written guidelines for handling a
specific situation.

Standard operating procedures (SOPs) are simply a record of a successful method of accomplishing a particular objective. Procedures are used every day of our lives. How you tie your shoes, the way you drive your car, and the recipes you use for cooking are examples of informal procedures. When the need for safety, reliability, or exact replication of the method becomes more critical, then formal, written procedures address those needs. The one thing that SOPs are not, is the exact method of always achieving an objective. In reality SOPs are guidelines based on a particular set of circumstances. The circumstances may be fairly broad, but at some point the situation you face may not fit within those circumstances.

An example is a procedure that requires a ladder to be thrown to a second-floor, street-side window of all residential buildings. The original intent of this SOP may have been to provide a means of escape for firefighters at private dwelling fires; however, since this was written, the community has changed and there are now five-story and taller residential buildings in the community. What would be the purpose of automatically throwing a ladder to the second-floor window of a five-story residential building? In reality, this is not the circumstance for which the SOP was written, but when blindly following it, the rescuer can create problems. To prevent this, people must understand the background and purpose of an SOP—as the procedure is being developed it must be reviewed for applicability and intent and the people who will use it must be trained to understand the background information. The user must understand the limits of the SOP and have enough knowledge to determine when to apply and when to deviate from it. That knowledge comes from training and experience and people should be taught to apply an SOP only after they look at the situation and see that it fits the parameters of the SOP. Blindly following a procedure shows a lack of understanding and poor application of knowledge and training.

WRITTEN SOPs

Written SOPs are formal documents that give an exact record which can be passed from person to person without missing a critical step. The more critical the need for exact replication of a procedure, the more critical the need for a formal written SOP. A written confined-space entry program is a good example of a written SOP. Developing a written entry program means that people will enter and work in confined spaces according to certain protocols. Entry will be accomplished only after certain steps are taken, and if there is a hazard present then it will be addressed. Although that is an oversimplification of the program, you can see that certain crucial items must be addressed.

However, how many of those items are different for every confined-space entry? The requirements to conduct atmospheric testing, review lockout/tagout needs, and provide an attendant, entrant, and entry supervisor are generally similar. That similarity provides a basis for the

SOP, addresses the problems, and eliminates the need to reinvent the wheel every time a confined-space entry must be made. An SOP also allows you to apply a set solution to a specific problem. So in addition to providing a set of guidelines to follow for a specific situation, an SOP can be the basis for starting a rescue operation.

You would begin by determining those items that are routine for most confined-space rescues. From those items that are shared from incident to incident, you would begin your SOP. Recall in chapter 9 the nine-step process for confined-space rescue. You could use those nine steps to develop the most basic SOP, start your operation, and then expand it from there.

Of course this would be a very basic SOP and you might find that it does not work as well as you desire. Expand the items that need to be addressed. If you think that the first step of establishing command needs better definition, use the SOP to expand it. You may think that a safety officer must be appointed at every confined-space rescue. Then specify, when command is established, that the incident commander must also appoint a competent individual as safety officer, but simply mentioning the need for a safety officer does not define that person's role. Is the safety officer responsible for scene safety or for safety at the confined space? The SOP is a means of communicating what needs to be done, and if the situation fits within its limits, you will achieve a certain level of performance where the routine items will have been accomplished. You can go through each of the nine steps and develop a basic SOP that meets your needs. Meeting your needs is important because of the limitations that you face based on your equipment, personnel, resources, and types of confined-space rescues that you expect to encounter. When you come across a rescue situation that does not fit within the SOP, do not abandon the SOP. Either determine what you can use from your existing one or have an alternative built into it. Having an alternative in the SOP is simply an expansion of it and shows that you are trying to anticipate those situations which are beyond your abilities.

SOPs are documents that are meant to serve as basic guidelines. They can be fairly complex and thorough, and should be studied in advance and not as you arrive at the scene. SOPs are also valuable training tools, because they ensure that everyone involved has an idea of how the operation should begin and what roles they may play in the rescue. Written, formal SOPs also provide that everyone attending the training will receive the same information and will be able to access that information in the future, which is no simple task. Think of the number of different things that you were taught informally. Now try to explain one to another person and have that person explain it to still another person. Consider the simple process of tying your shoes. You begin by pulling the laces tight and then tying an overhand knot. After you complete the overhand knot, you create a bight in one end of the lace and wrap the other lace around it. As you wrap the loose end around the bight you form a bight in the loose end and pull it through the knot. Sounds simple enough, until you try to teach a child to tie her shoes. Now imagine if you expect that child to be able to teach another child. If you tie your shoes improperly what is the worst that can happen? You might trip over your laces. What about those times where the life of emergency responders and the victim depend on being able to properly follow a procedure? Would you like to leave the crucial steps to memory? As stated earlier, written SOPs provide a point to which you can return continuously to provide training and procedures that can be replicated.

The SOP should be written clearly and should identify the type of situation for which it is intended, as well as who is expected to adhere to it. In the event that you had different levels of training and roles for personnel, the SOP should have a section that addresses those differences in training and roles. Many emergency responders should be familiar with this process from a hazardous materials response when different levels of training (awareness, operations, technician) and different types of actions (defensive or offensive) can be taken. If the levels of training are so different that some personnel will have a very limited role while others will have an expanded role, then you may have to write an SOP that starts the operation—if it is discovered by responders with limited training—and then write separate SOPs for the rescue team. However you decide to write the SOPs, write only those that address your needs. Remember, you are looking for a guideline that will build on successful methods of accomplishing the objectives set out for rescue operations. No SOP can address every situation, but there are enough similarities between incidents that you can find a common thread to use. At times the common thread will be a strong, integral part of your operation. At other times, the thread will be weak and tenuous at best and will be nothing more than a starting point.

CHECKLISTS

Standard operating procedures have tremendous value in that they formalize your operations, but consider an SOP that is five or ten pages long. Although you have trained with the SOP and are familiar with it, you cannot hope to remember it verbatim. How do you apply that SOP and feel confident that you have thoroughly covered each critical step and in the proper order of completion? The answer is to develop a checklist that reflects your SOP, as shown in Figure 10-1. Checklists coincide with SOPs in that they briefly identify each critical point and then allow you to record whether you were able to accomplish them or if they do not apply in the particular circumstances you have. It is much simpler to look at a one- or two-page checklist than it is to look at a five- or ten-page SOP. Checklists are also valuable for training in that you can use them to evaluate a training evolution based on an SOP. When the students have had the information presented to them and a chance to apply the information, use the checklist to identify that each required step has been successfully completed. If the students do not complete the required steps or are all having difficulty completing a particular step, you may have identified a serious problem in either the capabilities of your personnel,

FIGURE 10-1

Confined-Space Rescue Checklist

Step 1—Establish command

Command established and identified	☐ YES	☐ NO

Command is: _____

Command has been transferred to: _____

Step 2—Identify the type of rescue problem (choose only one)

Confined-space rescue	☐ YES	☐ NO
Rope rescue	☐ YES	☐ NO
Hazardous materials incident with rescue	☐ YES	☐ NO
Other nonfire rescue	☐ YES	☐ NO

Step 3—Perform hazard and risk assessment

Confined-space entry permit present	☐ YES	☐ NO	☐ N/A
If "NO" can processes and/or potential hazards be identified?	☐ YES	☐ NO	☐ N/A

Date and Time Issued: _____ Date and Time Expires: _____

Job site/space ID correct	☐ YES	☐ NO	☐ N/A
Job supervisor identified and present	☐ YES	☐ NO	☐ N/A
Equipment to be worked on identified	☐ YES	☐ NO	☐ N/A
Work being performed identified	☐ YES	☐ NO	☐ N/A

List: _____

_____

_____

All on-site personnel accounted for (include victims)	☐ YES	☐ NO

Number of victims: _____ Number of on-site personnel_____

Location of victims: _____

Required MSDSs at site and available	☐ YES	☐ NO	☐ N/A

Atmospheric Checks: Check once before ventilation, continuously monitor, and then log at 15-minute intervals after ventilation.

Oxygen: <19.5% oxygen deficient: > 23.5% oxygen enriched, Explosive: >10% LFL potentially explosive, Toxic see TLV or TWA.

Time	Oxygen—%	Explosive %LFL	Toxic—PPM

Identity of Toxic Materials _____

FIGURE 10-1 cont'd

Step 4—Identify rescue objectives

Victim rescue:	☐ YES	☐ NO
Can you identify defensive operations?	☐ YES	☐ NO
Can you identify offensive operations?	☐ YES	☐ NO
Victim recovery	☐ YES	☐ NO

Record your objectives:

_____

_____

_____

Step 5—Identify resource needs to support rescue objectives

_____

_____

_____

_____

Direct reading gas monitor—tested	☐ YES	☐ NO	☐ N/A
Safety harnesses and lifelines for entry and standby persons	☐ YES	☐ NO	☐ N/A
Hoisting equipment	☐ YES	☐ NO	☐ N/A
Powered communications	☐ YES	☐ NO	☐ N/A
SCBAs or SARs for entry and standby persons	☐ YES	☐ NO	☐ N/A
Protective clothing	☐ YES	☐ NO	☐ N/A
All electric equipment low voltage and intrinsically safe and nonsparking tools	☐ YES	☐ NO	☐ N/A

Step 6—Develop an action plan

_____

_____

_____

_____

Ventilation

Ventilation started:	☐ YES	☐ NO	☐ N/A
Positive pressure mechanical	☐ YES	☐ NO	☐ N/A
Natural ventilation only	☐ YES	☐ NO	☐ N/A

Communication procedures:_____

Rescue Procedures

Rescue procedures identified and communicated to rescue team for:			
Victim(s)	☐ YES	☐ NO	☐ N/A
Rescuers	☐ YES	☐ NO	☐ N/A

Entry, standby, and backup persons:	☐ YES	☐ NO	☐ N/A
Completed required training?	☐ YES	☐ NO	☐ N/A
Is it current?	☐ YES	☐ NO	☐ N/A

FIGURE **10-1** cont'd

Step 7—Implement the action plan

The following has been reviewed and is acceptable for rescue operations to begin:

Lockout/deenergize/tagout	☐ YES	☐ NO	☐ N/A
Line(s) broken-capped-blanked	☐ YES	☐ NO	☐ N/A
Area secured (post and flag)	☐ YES	☐ NO	☐ N/A
Breathing apparatus—SCBA or SAR	☐ YES	☐ NO	☐ N/A
Full body harness w/D-ring or other acceptable harness	☐ YES	☐ NO	☐ N/A
Emergency escape retrieval equipment set up and checked	☐ YES	☐ NO	☐ N/A
Lifelines	☐ YES	☐ NO	☐ N/A

Respiratory Equipment Time Log

Rescue Team 1		Rescue Team 2	
Name_____ SCBA or SAR (circle) Pressure_____ On air time_____ Air duration_____ Recall time*_____ Off air time_____	Name_____ SCBA or SAR (circle) Pressure_____ On air time_____ Air duration_____ Recall time*_____ Off air time_____	Name_____ SCBA or SAR (circle) Pressure_____ On air time_____ Air duration_____ Recall time*_____ Off air time_____	Name_____ SCBA or SAR (circle) Pressure_____ On air time_____ Air duration_____ Recall time*_____ Off air time_____

*Recall time is on air time plus air duration time minus a minimum of five minutes for time to exit the space. For longer exit times, subtract more time from the combined air time and air duration time.

Step 8—Evaluate the effectiveness of the action plan

Plan is proceeding as expected	☐ YES	☐ NO
Plan is not proceeding as expected	☐ YES	☐ NO
Minor changes needed?	☐ YES	☐ NO
Major changes needed?	☐ YES	☐ NO
Changes communicated and effective?	☐ YES	☐ NO
Does the operation need to be stopped to make changes?	☐ YES	☐ NO

Changes to action plan: _____

_____

_____

Step 9—Terminate the incident
Check that all personnel are accounted for.
Retrieve, recover, and maintain equipment as needed.
Prepare incident reports.
Hold a debriefing and critique.
Evaluate operations and their effectiveness.
Consider if remedial actions are needed.

equipment, SOP, or training methods. A good example is if you were training personnel in the use of SCBA. If each of the students failed to properly turn on the air supply or could not control the air supply once it had been turned on, what would be the problem? Would it be a failure of the SOP, the equipment, or your personnel? Regardless of the cause of the problem, you now know where the problem is and can make adjustments as needed.

■ SUMMARY

Standard operating procedures formalize the basic operations that you must use for different types of incidents. The SOP is a guide to be used when the conditions of the emergency fit the conditions that the SOP was developed for. No SOP will fit perfectly at every incident and personnel must understand the basis and intent of it. Do not write an SOP just for the sake of writing one. When needed, you can develop checklists from your SOPs. Checklists can be used as a training tool and at emergencies. Checklists simplify the application of an SOP and are intended as a means of allowing the user to follow one accurately.

■ REVIEW QUESTIONS

1. Using the nine-step process in chapter 9, identify the minimal SOP that you would develop for response activities at a confined-space rescue.

2. Taking the information you developed in question 1, identify those areas that would merit development of expanded SOPs. Choose one area and develop an SOP to address that area.

3. Based on the SOP you developed, create a checklist to support it. Explain how you would use the checklist for training and at an emergency.

4. You have a standard operating procedure for a specific type of emergency response; however, if you follow the SOP, then this incident will only get worse and endanger more people. You should _____.
   a. Stick with the SOP to avoid trouble
   b. Modify the SOP so that it fits all emergencies
   c. Call for help so that you can have someone else figure out how to apply the SOP
   d. Accept that this incident is outside of the scope of the SOP and develop a plan for handling the incident

5. Checklists briefly identify each critical point in an SOP; therefore, you can effectively use a checklist before you have been trained in the SOP. True or False?

CHAPTER

11 | Rescue Equipment

OBJECTIVES

After completing this chapter, the student should be able to recognize the following types of equipment:

- fall-arresting devices
- Class I harnesses
- Class II harnesses
- Class III harnesses
- tripods
- retrieval equipment

define the following terms:

- impact load
- static load
- working load
- axial load
- eccentric load

and explain the proper loading of a tripod.

INTRODUCTION

When performing any type of rescue a certain amount of risk is present. That risk may affect the victim, the rescuers, or both groups of people. For most emergency responders this risk is simply accepted as a part of the job they perform. Without realizing it, rescuers reduce, control, or manage the risks they face. The control can be as simple as the use of firefighters' protective clothing, the use of latex gloves for EMS work, or the use of a bullet-proof vest by a police officer. This equipment is part of the protection from known risks faced in day-to-day emergencies. Many other risks are managed in less obvious ways and, in fact, may be so well managed that they may not even be recognized by most emergency responders. Examples of such risks are the requirements for periodic hydrostatic testing of compressed gas cylinders (SCBA cylinders and oxygen cylinders), the specific performance requirements for tires for the response vehicles, the fire hose, and the performance of biohazard protective equipment. But what happens when we face an emergency that is new or unique to us? What can we do to identify the risks? How much time do we have to control the risks and still successfully rescue the victim?

It has been said that there are no emergency situations that require advanced thinking, only a more skillful combination of basic concepts—an oversimplification, perhaps, but many basic details are run in the background of emergency operations. Among these basic concepts is the idea that equipment must be designed to withstand a certain load and be used within certain limits. By staying within those limits, the equipment remains reliable and thus it performs satisfactorily. When you violate those limits, you introduce an unacceptable level of risk. It is possible that you will exceed the limits of equipment and still succeed, but why would you knowingly accept those risks and endanger emergency personnel and victims? Were you pressured to rescue a victim and therefore chose to take on a greater risk? Were you being reckless? Or were you acting outside the limits in ignorance?

Most people think that assuming a greater share of the risk is heroic—you place your life at greater risk to save the life of another person. This act is noble, but if you chose to be reckless (and, hopefully, no one was injured or killed), at the very least think of how people would look at your judgment in future situations. If your excuse is ignorance, how valid is that, because you should know how to use your equipment properly. The level of training and education of personnel is a limiting factor. Lower levels of training should be matched to lowered expectations of the ability of personnel to handle a rescue incident. If you do not know or fully understand the basics, then you cannot address even basic rescue situations. At times you will know the basics, but not how to operate a specific piece of equipment, including how it was designed and intended to be used. This chapter will begin to address such an issue. It is not a substitute for learning the specifics of your equipment, but rather a place to begin to understand how different equipment is designed to be used in different ways.

TYPES OF LOADS

Equipment used to lift rescuers, victims, and other equipment is designed to carry a specific load, but not all loads are the same. Depending on how the force of the load is applied, the effects can vary greatly. The initial force to reckon with is the weight of the person or object (load) being supported. The weight is the force being applied by the pull of gravity. If the load is in motion then be concerned about the effects of acceleration and whether the load is being pushed or pulled in a horizontal or vertical direction.

The different types of loads include **static loads, impact loads, working loads, axial loads,** and **eccentric loads.** To understand these loads, you will have to understand how each is applied.

Static loads are applied and remain in the same position and location, as shown in Figure 11-1. Typically, the forces of a static load are applied in only one plane (horizontal or vertical). Examples of static loads are forces applied to a harness or life safety rope during testing, a building sitting on its foundation, or a tank sitting on its supports. The object is not moving and is not expected to move (unless under failure). For rescue purposes, most loads that you will worry about in the field are not static loads, because your loads and forces will be moving. Static loads are most valuable to us when we specify equipment and the standard requires testing at specific static loads.

Impact loads are more critical to rescues because they can greatly exaggerate the force being applied. When a load (rescuer or victim) is in motion and then brought into contact with another object (stationary or moving) the moving object will give up some of its energy to the other object, as shown in Figure 11-2. If the energy is given up in a very short time, the force will be greater than just the weight of the object due to the effects of acceleration. If, however, the energy is applied slowly, the effects will not be as great. An impact load is the application of a force from at least one object being in motion and will vary as to the effects based on the forces created by the motion. An example of an impact load is a person walking on a board. If the board is supported on each end and the person slowly lowers his weight onto the board, it might flex and bend but still support his weight, as shown in Figure 11-3. This is an impact load, but one that is applied slowly and within the limits of the board. However, if someone were to take that same board and jump onto it while

▶ **static load**

a load applied when the load is at rest.

▶ **impact loads**

a load applied in a very short duration so as to include the effects of acceleration in the load.

▶ **working loads**

the maximum weight that a rope is expected to support.

▶ **axial loads**

a load transmitted through the axis of its supporting device.

▶ **eccentric loads**

a load applied so that the force of the load is off center of the supports carrying the load.

FIGURE 11-1

Static load—a static load is applied in only one direction and the whole system is at rest.

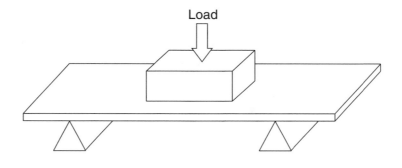

it was supported on each end, the expected result would be the breaking of the board, as shown in Figure 11-4, due to the speed with which the force was applied to the board and the fact that the acceleration greatly multiplied the force of the load. Impact loads occur frequently during confined-space rescues and although they should be controlled by using the equipment carefully, they may occur unintentionally. Impact loads may be applied in several directions (up, down, left, right, front, or back) at one time due to the movement of the load. Because of impact loads, equipment used to support hu-

FIGURE

An impact load is created when a load that is in motion is applied to the support.

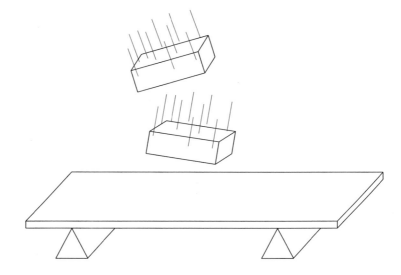

FIGURE

During impact loading, the load is in motion, and the acceleration increases the effect of the load.

FIGURE 11-4

The result of an impact load can be great enough to cause failure of the support.

man life must be able to withstand reasonably expected weights. It is not unusual for lifelines to require a minimum breaking strength of 4,500 foot pounds (lbf) for a one-person rope and 9,000 lbf for a two-person rope.

Working loads are expected to be applied to equipment during its use. The maximum working load is the maximum weight that is expected to be supported by the equipment.

An axial load refers to the direction that a load is carried. *Axial* simply means "moving about the axis." The axis in this instance is the centerline of a load-bearing point. Imagine that you are looking straight down on a tripod from above. As the victim is being raised, the direction of pull on the line supporting the victim is straight up and down, as shown in Figure 11-5. With the direction of pull in this position, the load is applied along the axis of the tripod. In turn, the load is transferred to the ground equally between the legs of the tripod and along the axis of each leg to the ground, as shown in Figure 11-6. This is how the tripod was designed to be used and you would not expect it to tip or be pulled over. If, however, the load was not transferred along the axis, but rather at an angle to the axis or off center of the axis, as shown in Figure 11-7, this load could cause the tripod to fail.

A load that is applied off center of the axis is an eccentric load, which can cause collapse or twisting of a support. As you set up your equipment for rescue, consider the direction of pull that you will be applying to any load and whether it will affect the stability of the point where it is being applied.

FIGURE 11-5

An axial load is applied in the same plane as the axis of the support.

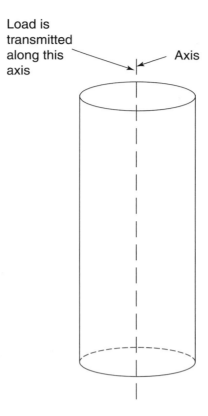

FIGURE 11-6

Even though a tripod has three legs to carry the load to the ground, there is still an axis for the entire tripod. Additionally, each leg of the tripod also has an axis.

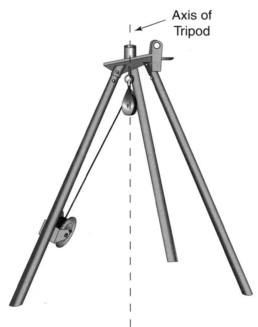

EQUIPMENT BASICS

Earlier in the book and in this chapter we discussed the topic of standards. Standards vary by organization as to the legal weight that they carry. Certain standards that have been developed and adopted by regulatory agencies, as shown in Figure 11-8, such as OSHA and CALOSHA, have the force of law. Other standards such

FIGURE 11-7

An eccentric load is one that is applied off center. The eccentric load can cause a failure of the support.

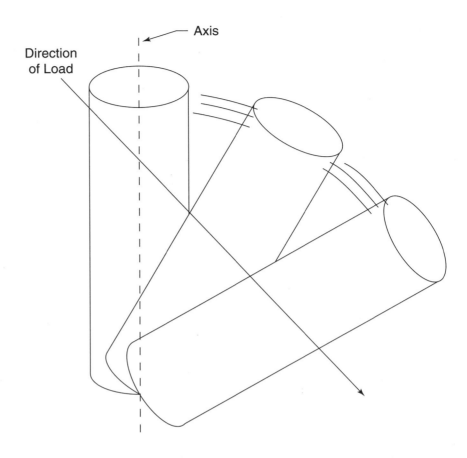

FIGURE 11-8

There are a variety of different standards which affect how different types of equipment are designed, manufactured and used.

as NFPA and ANSI are typically consensus standards and do not have the force of law unless they are specifically adopted or referenced by a government agency. Because organizations such as NFPA and ANSI are nationally recognized and they have very specific procedures regarding the development and adoption of a standard by the organization, these consensus standards may be used in court cases and may in fact be enforced as a result of a court case that creates case law. At the very least, consensus standards are valuable tools which will assist you in selecting and specifying equipment that will perform within certain limits and in a certain manner. You will not have to design your own equipment each time you need to purchase some and you can hold all of the vendors wanting to sell you equipment to the same standard; but you must understand the standards and how they apply to the equipment you are going to purchase.

You may often hear that a piece of equipment is OSHA certified or NFPA certified, but that is incorrect. Most if not all of the nationally recognized standards organizations do not test for compliance with their standards. The standard is only the criteria to which a piece of equipment must be designed and manufactured. It is the responsibility of the manufacturer to use a recognized third-party testing laboratory to determine if the equipment meets the requirements of the standard. Usually the standard contains or references a specific procedure that must be followed for testing the equipment for compliance. The purpose of these specific testing requirements is to allow the test to be performed in the same manner each time and therefore be able to achieve the same results each time for identical equipment. If the equipment is tested as required and meets the standard, it is then labeled as meeting the particular standard, as shown in Figure 11-9. Additionally, the testing laboratory is required to document that the equipment has met the requirements of the standard and the manufacturer should be required to provide that documentation to you when you purchase equipment.

Meeting standards is an expensive investment for a manufacturer. It is something that the manufacturer can be proud of because it sets its product off from other equipment. Occasionally, however, some manufacturers try to avoid a particular standard by outright fraud, such as using a false label or referencing another standard from the same standard organization and simply stating that it is "compliant." Such was the case when **Personal Alert Safety Systems (PASS)** units were first introduced to the fire service. A PASS unit is a battery-powered alarm for firefighters who may become lost, unconscious, or trapped during a fire. The NFPA had a specific standard for PASS units—NFPA 1982. One company that was selling a PASS unit had not yet passed the testing procedure according to NFPA 1982. In its haste to get the product to market, this company said that its PASS unit was NFPA compliant, but then referenced another NFPA standard that was designed to cover electrical equipment in hazardous locations. A clever ploy if the buyer did not know which NFPA standard was the correct standard to reference. Fortunately, the NFPA was made aware of the problem and they forced the company to change its advertising.

Standards, like any other document, can become outdated, so organizations periodically update them. It is important that when you purchase equipment, you reference the current version of the standard. The frequency with which standards are updated varies with the organization that produces it and you can easily call the

---

| **NOTE:** You may often hear that a piece of equipment is OSHA-certified or NFPA-certified, but that is incorrect.

▶ **Personal Alert Safety Systems (PASS)**

a device, typically worn by firefighters, intended to provide a means of automatically sounding an alarm when the wearer remains motionless for more than 30 seconds or manually by the wearer if in need of assistance.

FIGURE 11-9

This label, on a harness, shows the standards that the harness is designed to meet, the manufacturer, lot number, and model number.

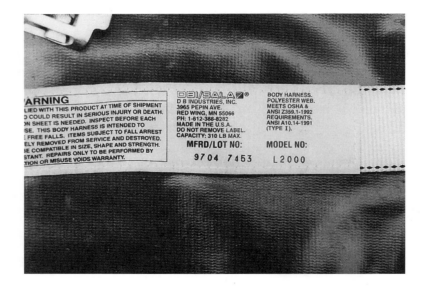

▶ **nondestructive test**

a method of testing in which an item is tested within specific parameters to determine if it can meet the requirements of the test and recover within acceptable limits.

▶ **destructive test**

a means of testing during which the test item is tested to failure.

organization to find out the most current standard. If you have older equipment that was purchased under an earlier version, it is generally still usable, but you should check the new version of the standard to see what changes have been made. Newer equipment may have higher load ratings or be used in different ways. If you see a new piece of equipment used in a particular way, it does not automatically mean that you can use your existing equipment in the same way.

Not all standards organizations are in agreement when they create standards that cover the same or similar topics. Groups such as NFPA and ANSI try to resolve conflicts between their standards, but at times this resolution is fruitless. In that case, you need to look at the standard, the organization, and the group of people who drew up the standard and make a decision on which to use. Consider also who is the intended user of the equipment outlined in the standard.

Standards are of value to us not only for the purchase of equipment, but also for its maintenance, as shown in Figure 11-10. Standards may often contain or reference information on how to maintain the equipment. The standard may also provide for specific tests that can be performed to ensure that the equipment still meets the performance requirements of the standard. Certain standards require testing—by using either **nondestructive tests** or **destructive tests.** Testing for periodic maintenance does not typically include destructive testing. Nondestructive tests allow you to test the equipment without destroying the equipment. Examples of nondestructive tests include hydrostatic tests for air cylinders and annual hose tests for fire hose. In these instances the equipment is tested and then returned to service if it passes the test. Nondestructive testing follows a procedure, but the complexity of the testing procedure may be simple enough that it can be done locally without a complicated setup (hose testing is a

FIGURE 11-10

A sample inspection and maintenance log. You should consult the manufacturer for their specific recommendations as well as recognized standards.

9.0 DETAILED INSPECTION & MAINTENANCE LOG:

SERIAL NUMBER: _____

MODEL NUMBER: _____

DATE PURCHASED: _____

INSPECTION DATE	INSPECTION ITEMS NOTED	CORRECTIVE ACTION TAKEN	MAINTENANCE PERFORMED
Approved By: _____			
Approved By: _____			
Approved By: _____			
Approved By: _____			
Approved By: _____			
Approved By: _____			
Approved By: _____			
Approved By: _____			
Approved By: _____			
Approved By: _____			

good example). Destructive testing, on the other hand, takes a piece of equipment and tests it until it fails. This destructive testing is of value in that a representative sample of the equipment is tested—not all of the equipment—and it may be the only method available to provide a reliable test for that particular piece of equipment. Unfortunately destructive testing causes the loss of equipment and is not often used outside of a manufacturer's testing facility or a third-party laboratory. Because the use of destructive testing typically requires specific test equipment it is usually done in a laboratory to ensure the ability to replicate near identical results from testing similar items.

HARNESSES

When we think of confined-space rescue, we may easily think of the harnesses that are used by both the rescuer and victims. Depending on how these harnesses were designed and who is the intended wearer, a variety of different harnesses are available for use. You should, however, only select harnesses that are designed for use in rescue work and meet the requirements of NFPA 1983 and ANSI A101.4. NFPA classifies harnesses in three categories: Class I harnesses, Class II harnesses, and Class III harnesses.

Class I harnesses are designed to go around the waist and the thighs or under the buttocks with the intention that they will be used for emergency escape and will only support one person, as shown in Figure 11-11. Class I harnesses should not be used during confined-space rescue when it is necessary or expected that a person will have to be raised or lowered using the harness. These harnesses should not be confused with ladder belts which are intended to secure people to a stationary object where they will remain standing in an upright position, as shown in Figure 11-12.

Class II harnesses consist of a waist belt with straps around the legs and buttocks that create a seat for the wearer and are expected to carry a two-person load, as shown in Figure 11-13. Class II harnesses

FIGURE 11-11

This is a class I harness. It is designed to support a single person. All harnesses should be clearly marked as to their class.

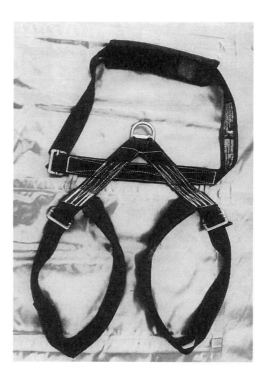

are worn when the person can be lowered or raised and is expected to remain upright or nearly upright. Class II harnesses are meant to allow the wearer to be supported by the seat and do not cause the back to absorb the forces applied during lowering or raising. They are not designed to be used when the wearer may be inverted or rotated perpendicular to the ground. Once the wearer of a Class II harness reaches an angle that is parallel to the ground, he may slip out of it and fall. Class II harnesses have limited use for confined-space rescue.

Class III harnesses not only include a waist belt and seat, but also provide straps that go from the waist area up the front to the shoulders and back down the rear of the body, and are expected to carry a two-person load, as shown in Figure 11-14. Class III harnesses provide protection from falling out of the harness if the wearer becomes inverted, but depending on the location of the D-ring connectors located on the harness, they also allow the wearer to be lifted by an attachment at the shoulders, center of the upper back or midchest, or the sides and center of the waist. This type of harness provides a high

FIGURE 11-12

The ladder belt worn by this firefighter should not be used to support a person while raising or lowering on a rope, cable, or other support.

FIGURE 11-13

A Class II harness. Class II harnesses are designed to support a two-person load and look very similar to a Class I harness. Only by consulting the marking on the harness can you tell if it is a Class I or Class II harness.

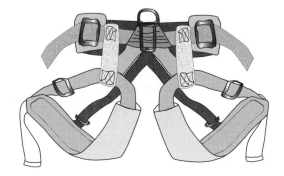

FIGURE 11-14

A Class III harness. This harness not only supports a two-person load, but it also protects a wearer from falling out if inverted.

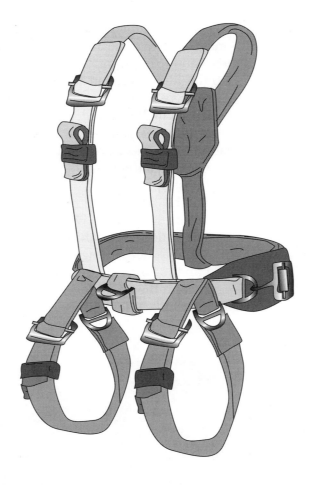

degree of adaptability for use at a confined-space rescue. Not only is it possible to secure a wearer who might invert, but it is also possible to lift a wearer at the shoulders, chest, or back by attaching a line to the D-rings. The advantage is realized when the situation involves an extremely narrow opening or an opening that requires the wearer to have his hands free for manipulating tools or equipment—the wearer can now be raised or lowered through the opening with his hands free for repositioning. For example, you might enter an 18-inch opening while wearing an airline respirator and escape bottle. To enter this opening you must have at least one hand free to align and pass the escape bottle and airline through the opening. Additionally, most people would have to raise their hands over their head and grasp the line to pass through the opening. Attempting to do either of these maneuvers while wearing a Class II harness would pose the risk of falling backward as you let go of the rope with your hands. A class III harness would keep your body in a straight line and allow you to have both hands free.

In addition to the type of harness that you are using, you must be aware of other limitations to harnesses and the equipment associated with them. To begin, what type of material is the harness constructed of? Is the webbing polyester, nylon, or some other material? How long of a life span does the manufacturer recommend for the

SAFETY
You must be aware of other limitations to harnesses and the equipment associated with them.

FIGURE 11-15

The stitching pattern and number of stitches per inch in a harness are important considerations since they can affect the strength of the harness.

harness? Different materials behave differently to abrasion, sunlight, and water. If the harness has been contaminated with oils, gasoline, grease, or other chemicals, what will be the effect on the harness? You must initiate a documented inspection program for your harnesses (hopefully one recommended by the manufacturer) and keep track of them through an identification system. One of the biggest problems that you may face when a harness is suspected of being damaged or no longer serviceable is that you cannot test it in a nondestructive manner. Aside from visual inspections, it is very difficult, if not impossible, to test a harness without destroying it. If you attempt to load test the harness, you will wind up loading the harness until it fails.

Other features of harness construction to be aware of include the stitching of the harness (number of stitches per inch and the pattern), as shown in Figure 11-15. Every time a needle is passed through the fabric of a harness it will break a certain number of fibers within the material. The more stitches per inch, the more broken fibers in a harness and thus a weakening of the belt. Some manufacturers recommend five to seven stitches per inch. Not only is the number of stitches per inch important, but also the stitch pattern. A stitch pattern should be able to resist the forces of the load being applied to the harness and spread the load over a larger area of the belt. Manufacturers today use computer-controlled sewing machines. A consistent stitch pattern and a consistent number of stitches per inch should not be a problem for a quality harness.

Although they are not harnesses, **wristlets,** as shown in Figure 11-16, may be used in specific circumstances when it is not possible to use a harness. Wristlets may also be used around the ankles for a horizontal entry when the victim or rescuer could not use a harness. Whereas the webbing of a harness is required to have a minimum breaking strength of 6,000 pounds, wristlets are more difficult to categorize. Because wristlets carry the load through joints within the body and do not support it, the safety factor is 3:1. This factor means

SAFETY

You must initiate a documented inspection program for your harnesses and keep track of them through an identification system.

► **wristlets**

straps designed to be placed around the wrists of a person to allow him to be raised or lowered through a vertical opening while hanging from the straps.

FIGURE 11-16

Wristlets that can be used to raise
or lower a person.

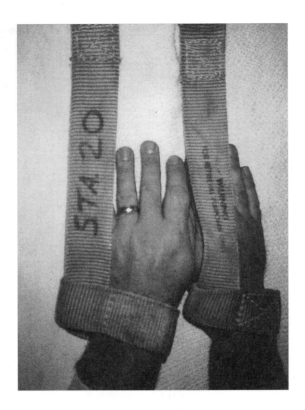

that if you intend to lift a 300-pound load, the wristlets must have a
minimum breaking strength of 900 pounds. The breaking strength of
wristlets will vary and you must be aware of the strength, but few if
any wristlets will have a breaking strength greater than 5,000 pounds.

Additionally, you must consider how wristlets are intended to
be used. Attaching the retrieval equipment to a person's wrists and
then lifting by raising his hands over his head places all of the strain
on the shoulder joints and muscles. Depending on the size of the person being raised or lowered this way, it can be painful and may cause
injury. If you intend to use wristlets (and you should consider them
as part of your equipment), look at the wristlets and how they attach,
how comfortable they are to wear around the wrist, and if they can
be adjusted to compensate for body size. You may also want to consider wristlets that are large enough to go around a rescuer's ankles.
When the situation involves a very narrow pipe, and a rescuer must
crawl in horizontally, wristlets can be attached to the ankles and line
attached to wristlets to haul a rescuer back out when he cannot crawl
backwards or the rescuer can only grab the victim to drag her out.
Wristlets are a tool in that a victim or rescuer with no other means
out of a confined space can be attached to retrieval equipment and
hauled out. Be aware of the potential for injury and the limits of the
equipment, and be sure that you include wristlets in your maintenance and inspection program.

The inspection program that is in place should document the
history of your harnesses and equipment, and reflect the manufacturer's recommendations for an inspection program, which means
that the inspection you perform should include all items that the
manufacturer identifies as important to maintaining the reliability of

**NOTE: The inspection pro-
gram that is in place should
document the history of your
harnesses and equipment
and reflect the manufacturer's
recommendations.**

FIGURE **11-17**

The safety tabs shown on this harness are an integral part of the harness and must be inspected. The manufacturer of this harness puts these tabs in to show if the harness has been impact-loaded.

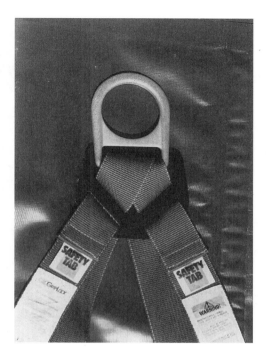

the harness, as shown in Figure 11-17. Fraying of the material, damaged stitching, chemical exposure, and fading of the fabric due to sunlight exposure should all be addressed in an inspection program. You must inspect harnesses periodically—as well as before and after each use. Inspecting the harnesses before use should involve a brief look to make sure they have not been damaged in storage. This before-use inspection should be a part of your SOPs for emergency response.

Any discussion of harnesses must include D-rings, O-rings, snap hooks, buckles, and other connecting hardware, as shown in Figure 11-18. The specific requirements for this hardware are interesting because the strength requirement for the harnesses require that the webbing has a minimum breaking strength of 6,000 pounds, whereas the breaking strength requirement for the connecting hardware is 5,000 pounds. This fact refers to the performance of the equipment that is presently available. You should be aware of this limitation in regard to the procedure used to test the harness for compliance with the standard. Some manufacturers test their harnesses with the connecting hardware in place, others do not. Ask your equipment vendor what testing procedures were used. The harness can be tested without the connecting hardware as part of the test and still meet the requirements of the standard. It is simply a matter of whether you are willing to accept this type of testing.

O-rings, D-rings, and other connecting hardware must also be inspected as part of your periodic, after-use, and before-use SOP. You should be inspecting every piece of equipment that will be used to support human life. The before-use inspection is the responsibility of everyone at the emergency, but you should consider assigning it as a specific task for the safety officer or other person to be sure that the inspection is performed before anyone is put on line.

FIGURE **11-18**

This D ring is built into the harness and must be inspected as a part of the harness. The strength of the D ring must be taken into consideration when using a harness.

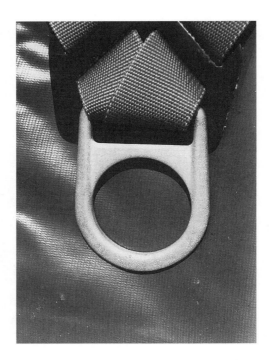

TRIPODS, QUAD PODS, AND OTHER LEGGED RESCUE EQUIPMENT

Tripods may very well be synonymous with confined-space rescue, but your choice of equipment is not limited to them. A tripod and other similar equipment provide an anchor point for lifting in a vertical position. For this section of the book, the word *tripod* will be used to describe all tripods and similar equipment. This equipment provides an anchor point for the lifting equipment or a change-of-direction pulley, but does not necessarily provide an anchor to the ground. When lifting with a tripod or similar equipment, you must watch how the equipment is loaded and make sure that all loads are axial loads to keep from tipping or collapsing the equipment. In addition to tripods, manufacturers are now making equipment with four legs and a davit arm (as shown in Figure 11-19), davit arms with U-shaped bases, a tripod designed to be bolted to the flange of a man way opening for a vertical confined space, attachment devices to allow rescuers to attach hoisting equipment to an overhead steel beam (as shown in Figure 11-20), and other specialized equipment for anchoring for an overhead lift.

The equipment that is available today varies greatly in its lifting capacity, use on sloped tank roofs or ground surface, number of lines that can be attached at one time, and height of lift. The selection and purchase of a tripod or other similar equipment should be made with a true picture of how you expect to use it, which requires a certain amount of preplanning. You must also have an idea of when the equipment you want to purchase will be inadequate for your rescue needs and whether you can adapt the available equipment or develop another means of rescue.

The lifting capacity of hoisting equipment can vary. At least one tripod manufacturer with built-in hoisting equipment reduces the

FIGURE 11-19

As confined-space rescue evolves, new equipment will be designed and built to improve operations. This is a Quad Pod®, which consists of four legs to support the load and a davit arm with a fall protection device attached.

FIGURE 11-20

A sling is being used to attach a change-of-direction pulley to the rung of an aerial ladder. This type of sling and other more specialized equipment is available to attach rescue equipment to beams, ladders, and other support points.

NOTE: Know the lifting capacity of your equipment.

allowed load on its equipment as the amount of rope being used gets longer. It is your responsibility to know the limitations of your equipment. Not all hoists can carry the same capacity. Know the lifting capacity of your equipment, because it can vary among models and manufacturers. In addition to a limitation on the capacity of the equipment are different manufacturers whose equipment can be used either on or off a tripod. If you purchase equipment that can be used off a tripod, be sure that you know how that hoisting equipment is meant to be used. Attaching the hoist incorrectly or at the wrong point can lead to a connection failure and thus can drop anyone who is connected to that device.

You may need to pay special attention to the usability of the tripod in some situations, such as when confronting a sloped tank roof with a smooth surface around the man way opening or a below-ground confined space where the entry is surrounded by loose soil. Many tripods and similar equipment have swiveling feet designed to provide traction on smooth surfaces or pointed feet on the reverse end to allow the foot to be pushed into soft ground for anchoring, as shown in Figure 11-21. It may become obvious through your preplanning that a regular tripod will not work. At this point, you may decide to acquire special equipment for a potential rescue at that particular confined space. Keep in mind the possibilities and probabilities. What can possibly occur may not have a very high degree of probability to occur. Prepare for what is probable and plan for what is possible.

Sloped surfaces for placing a tripod contain other hazards besides the footing. The method of loading the tripod is critical to the safety of people using it. On a sloped surface, two legs of a three-legged tripod should be placed on the same plane below the opening, as shown in Figure 11-22. In addition, the hoisting equipment should be placed on the uphill leg of the tripod to create a stable base below the opening and then load the weight to assist in anchoring the uphill portion of the tripod. Sloped surfaces also bring into play the importance of loading the tripod. You must imagine that between all legs of your tripod is an invisible line that is perpendicular to the head of the tripod and the earth. This line is the **axis** on which the load being hoisted out of the confined space must travel. If you do not maintain this axis perpendicular to both the earth and the head of the tripod, you may tip over the tripod. For devices with

▶ **axis**

the imaginary line passing through the center of a solid or plane

FIGURE 11-21

The foot of this tripod is designed to swivel so that it can be used flat on hard surfaces (as shown) or flipped up so that the pointed end can be pushed into soil and other soft materials.

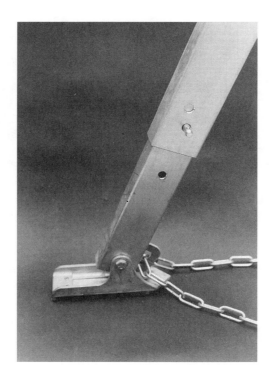

four legs, you will need to keep the base as close to level as possible because these devices transfer the load down the davit arm and then to the base. However, proper loading of the tripod does not end there. You must also carry the load down the legs of the tripod in an axial manner or you may collapse the legs or tip the tripod. The loads on the tripod must be spread between all three legs for the tripod to carry the full load it is designed to support. It may be possible to raise a load that is not applied axially to the legs, but you must find another anchor point, as shown in Figure 11-23, to maintain an axial load on the tripod and legs.

FIGURE 11-22

Tripods that are to be placed on a sloped surface should have two of the legs placed on the same plane downhill of the opening in order to provide the greatest stability.

FIGURE 11-23

The load shown in this drawing is being applied axially to the tripod by using a change-of-direction pulley between the legs of the tripod. If the change of direction pulley was outside the feet of the tripod, the load would not be axial, and the tripod would tip over.

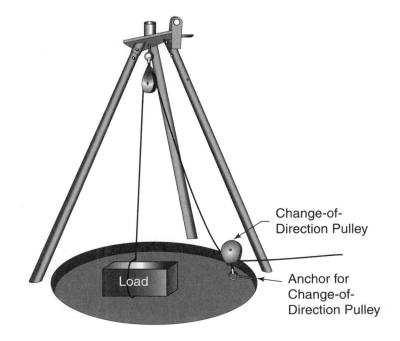

Change-of-Direction Pulley

Load

Anchor for Change-of-Direction Pulley

The locking device at the head of this tripod keeps the legs rigidly in place. If the tripod were tipped and there were no locks, the legs would pivot to the center of the tripod and the tripod would collapse.

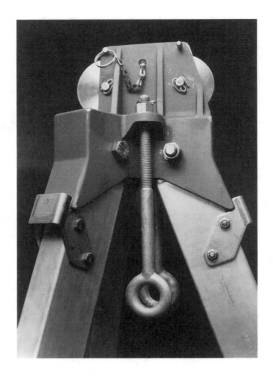

▶ **transformer retrieval support**

a device which is designed to be bolted to the flange around a manway opening for vertical entry.

Some tripod legs have a locking mechanism which locks the legs to the head of the tripod, as shown in Figure 11-24. If a tripod is tipped over, even slightly, the legs are not rigidly held in place. The weight is then on a single leg and the other two legs are free to swivel into the closed position. As you try to push the tripod back into position, the other two legs are not spread out and cannot take the load. With the legs closed, the tripod falls over and the person on the tripod is dropped. Locking the legs will not stop the tripod from tipping, but as you push this type of tripod back into position, the legs remain open and the tripod will have all three feet spread out on the ground. You should also consider how the legs are connected at the base of the tripod. Simple laws of physics want to spread the legs out and flatten them to the ground. Having some type of anchor such as a chain, as shown in Figure 11-25, between all of the legs of the tripod stops this problem from happening. If there is no attachment between the legs of the tripod, all of the forces trying to spread the legs apart are transferred to the head of the tripod and multiplied due to the lever effect of the load being transmitted up the legs.

As confined-space rescue continues to evolve and the equipment needs are refined, more versatile equipment will appear on the market. One of the most difficult places to put a tripod is on the sloped roof of a tank or between several narrow and close confined spaces. These types of rescue situations deserve preplanning (if you can find out in advance) and may warrant some permanent equipment mounted in the area. Recently, one tripod on the market was specifically designed for this type of situation. Called a **transformer retrieval support,** it is designed to be bolted directly to the flange of a manway opening, as shown in Figure 11-26. It may be the first of its kind, but it will not be the last.

For tripods that have height-adjustable legs, as shown in Figure 11-27, you must know what the manufacturer recommends in terms

FIGURE 11-25

Chains between the tripod feet keep the legs from spreading as a load is applied.

FIGURE 11-26

This transformer retrieval support is specifically designed and built to be bolted to the manway opening. It has a very specific use and is a very valuable device where the use of a tripod would be limited. (Photo courtesy of DBI/SALA.)

FIGURE 11-27

For tripods with adjustable legs, you must know how the performance of the tripod is affected by raising the legs.

FIGURE 11-28

The number of retrieval devices that can be attached to a tripod will have an impact on your rescue operations. This tripod has two devices attached.

of changes to its load-carrying ability. The longer the tripod leg, the less weight it can carry. Do not overload your tripod as you extend the legs. Know what effect the leg extension has on the tripod and stay within the manufacturer's limits.

The number of retrieval devices that can be attached to a tripod, as shown in Figure 11-28, will also affect your rescue capabilities. One obligation that any incident commander has is to protect the life safety of the emergency responders. Rescuers who must enter the confined space are taking on additional risk. The incident commander has the obligation to reduce and manage that risk. All persons who enter the confined space should be wearing a retrieval line so that they can be rescued from outside the space if they become unconscious or incapacitated while inside. Attaching retrieval lines to a rescue entrant speeds up the time it would take to rescue that person. Therefore, you need one retrieval line per rescuer. In addition to one line per rescuer, you must consider a safety line for all vertical rescues while a person is on the line and being raised or lowered, which means that you need one retrieval line per person and one safety line per lift, as shown in Figure 11-29. If you have two rescuers entering, you should have a minimum of three lines available—one for each rescuer's retrieval and one separate safety line for lifting—and three attachment points for hoisting on the tripod or other anchor point.

By managing who is on line you can send in two rescuers and still rescue the victim using only three lines. Both rescuers enter, one at a time, and detach the safety line (all rescue personnel in the space should keep the retrieval line attached). The victim is packaged and prepared for the lift. One rescuer is removed from the confined space, both lines from that rescuer are sent back into the confined space, and the remaining rescuer attaches the victim to the safety

FIGURE 11-29

The use of a safety line by rescuers should be mandatory. A safety line allows the rescuer to be protected in the event of a failure of the main line and allows the rescuer to remain attached to a retrieval line while working in the confined space.

▶ **Quad Pod™**

a brand name for a retrieval support device consisting of a davit arm centered over four legs which support the device.

line and available retrieval line. The victim is removed and the safety line is sent back in for the remaining rescuer. By managing the lines and how they are used, no rescuer is without a line for emergency retrieval, and no person is supported by only a single line. You not only have to worry about the failure of a rope or cable, but also about a failure of the mechanical parts of a retrieval device. Two lines minimize the effects of a failure.

Retrieval devices with four legs and a davit arm are not actual tripods. Depending on the manufacturer they might be called **Quad Pods™** or some other trade name. These devices must be loaded with as much care as a tripod. The use of a davit arm that extends between the feet of the device relies on the device being loaded properly. If the device is tilted or the load is not applied axially, it may tip over or be pulled away from where it has been set up, as shown in Figure 11-30. Depending on the features available, you must know what limitations the manufacturer has placed on the use of its equipment. Setting up a Quad Pod–like device on a sloped roof of a tank will be limited by the slope of the roof, the amount of adjustment available in the feet, and the type of surface on the tank roof. You should attempt to set up these devices so that at least two feet are on the downhill side of the opening and the hoisting equipment is uphill of the opening.

As discussed, the length of the tripod's legs affects the working limits of the tripod. You should also consider how high of a lift the tripod can give you, as shown in Figure 11-31. Once a victim or rescuer has been brought to the opening of the confined space, is there enough room to completely lift him out of the opening using the retrieval equipment? If a victim in a rescue stretcher is 6 feet long and the lifting height from the opening is just 6 feet or less, can you effectively get that person completely out of the opening using the retrieval equipment, or do you have to lift him as far as possible and then wrestle with the stretcher to get it clear of the opening? This problem may seem unimportant except that the victim may weight

FIGURE 11-30

Just as a tripod must be loaded axially, so must other devices. This illustration of a Quad Pod® shows how the load must be kept within the four legs of the device in order to prevent tipping or other failures.

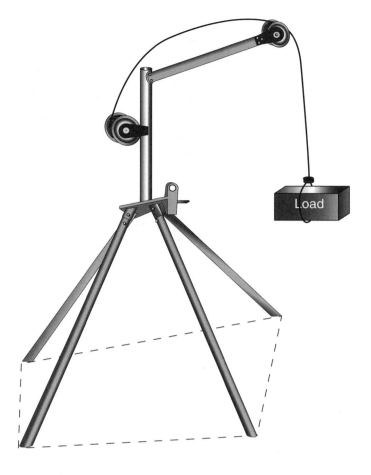

FIGURE 11-31

The height of the lift that a tripod or similar device can provide should be among your considerations when selecting the device.

300 pounds or be on the top of a very narrow, sloped roof tank or have some other problem that makes the limitations of the equipment a strategic factor.

HOISTING DEVICES AND FALL PROTECTION

There are many different ways to haul victims or rescuers out of a confined space. The solutions become a matter of efficiency, mechanical advantage, safety, and ability to learn to use the equipment while maintaining proficiency. The simplest equipment to use may be mechanical winches designed for confined-space rescue, but the simplest may not match the current circumstances. At other times, rope, pulleys, carabiners, and other types of rope equipment may be the best to use, but do your people know how to use them safely? This chapter will only briefly discuss the use of rope and other rope rescue equipment for confined-space rescue. The information needed to effectively learn rope rescue is beyond the scope of this book. The use of rope rescue equipment, techniques, and procedures for confined-space rescue (commonly called high-angle rescue) requires a separate course of study and demands a high level of training to maintain proficiency.

Retrieval winches are designed for hoisting people. These winches usually have stainless steel or galvanized steel cables attached for use as lifting devices to provide a mechanical advantage in raising and lowering people. (NOTE: Not all retrieval winches can be used for controlled lowering—some only offer a raising mechanism.) Other advantageous features include built-in fall protection, built-in handle brakes to prevent movement of the load when you release the handle, the ability to adjust the mechanical advantage provided, and a clutch mechanism (as shown in Figure 11-32) to prevent applying the force of the mechanical advantage on an entangled person. Disadvantages include limited lengths of the cables, inspection requirements that require some units to be returned to the manufacturer for periodic inspection and maintenance, cost per unit, the need to buy a separate unit for training use, and some manufacturers require that the unit be sent back after each rescue use for inspection. You need to look at the advantages, disadvantages, and your own needs to determine if this equipment is right for your purpose.

> **NOTE: You need to look at the advantages, disadvantages, and your own needs to determine if this equipment is right for your purpose.**

A winch that also provides a self-retracting safety line, as shown in Figure 11-32, can be a great advantage during confined-space rescue. If you had a limited number of team members available and had to maintain a safety line and a retrieval line, it would be difficult to maintain the correct amount of slack on the safety line while raising or lowering the person with the retrieval line. A self-retracting safety line provides that the line coming from the winch is free to extend or retract as fast as the wearer moves; however, if the wearer moves too quickly (accelerates such as during a fall) the line locks and stops the fall. Self-retracting winches limit a free fall to 18 inches or less, and during the raising or lowering operation these lines will not have to be tended by a person.

FIGURE 11-32

This retrieval device has a variety of features that makes it valuable for confined-space rescue. In addition to being useful in several different positions, it can be used as an untended safety line since it provides fall protection.

SAFETY
Use only double-locking snap hooks for rescue operations.

▶ **ANSI Standard A10.14-1991 (Requirements for Life Safety Belts, Harnesses, Lanyards, and Lifelines for Construction and Demolition Use)**

a specific ANSI standard that specifies the design, construction, maintenance, and testing of life safety belts and harnesses, among other items.

Winches and fall protection equipment often use snap hooks, as shown in Figure 11-33, to allow you to connect D-rings and O-rings to the life line. Snap hooks are available with either spring-loaded or double-locking closures, as shown in Figure 11-34. Use only double-locking snap hooks for rescue operations. You must match the snap hook to the minimum size attachment to which it will be connected. The throat opening of a snap hook that is too large can potentially allow the O-ring or D-ring to either side load the gate or slip under the keeper of the gate and slide out of the snap hook. In addition to matching the snap hook to the attachment, you must be certain that any load which is applied to the snap hook is along the major axis of the snap hook. The keeper of the snap hook is only tested to withstand a 350-pound side load. Loads greater than 350 pounds applied sideways to the keeper of the snap hook can deform the keeper and leave an opening where the keeper was. The **ANSI Standard A10.14-1991 (Requirements for Life Safety Belts, Harnesses, Lanyards, and Lifelines for Construction and Demolition Use)** recommends that for snap hooks with throat openings greater than 1 inch a locking feature be included in the design. Snap hooks must also be inspected and maintained. Follow the manufacturer's instructions and be sure to check for corrosion of the metal, distortion of any part of the snap hook and keeper, and the proper functioning of the swivels, springs, and other movable parts of the snap hook.

How you use the equipment is also important for ensuring the safety and reliability of hoisting equipment. Such equipment (tripods, harnesses, slings, winches, etc.) that is intended to be used for rescue purposes should not be used to raise or lower equipment, tools, or materials into or out of a confined space. Equipment intended for rescue use should be used only to support human life. Using the rescue equipment to move people into or out of a confined space during routine work practices is acceptable. However, raising

FIGURE 11-33

A snap hook for connecting O rings, D rings, and other equipment to retrieval equipment.

FIGURE 11-34

Match the snap hook to the size of the device to which it is to be connected. The snap hook on the left is a spring-loaded snap hook, while the one on the right is a double-locking snap hook. Double-locking snap hooks are the only type of snap hook that should be used to support human life.

SAFETY
Hoisting equipment that has already been overstressed by lifting tools and equipment can fail because of the demands of an emergency and injure or kill both the victims and rescuers.

or lowering equipment or tools can overload the tripod, winch, or other apparatus associated with the rescue equipment. This overload may not be visible or obvious, but when the equipment is used in a rescue it can fail. Equipment used for rescue purposes must perform to its design limits during an emergency, which explains why there are so many standards for rescue equipment. There must be that high degree of reliability and dependability so the equipment will not be overstressed and fail when it is needed most. Hoisting equipment that has already been overstressed by lifting tools and equipment can fail because of the demands of an emergency and injure or kill both the victims and rescuers. Some manufacturers make tripods and associated equipment with the claim that the equipment can be used for hoisting people and equipment. The problem is, did someone use the equipment with too large a load on it? For example, someone from your emergency organization borrowed the equipment to lift an engine out of a car. Not everyone understands the importance of this equipment and the proper use of it. This problem is difficult to manage and you must consider how much confidence you would have in that equipment if you were the person being supported by it. Manage the risks associated with the rescue problem—if you cannot be sure that the equipment has been used appropriately, can you be sure that you have not increased the risks?

ROPES AND ROPE EQUIPMENT

Using rope and rope equipment for confined-space rescue is fairly common, but for some emergency responders, it may be beyond their capabilities. Particular concerns about the use of rope include the rope and equipment and the proficiency that your emergency responders will have in using the equipment properly. Tying knots correctly is an invaluable skill, but one that can be lost if you do not practice. Using rope for confined-space rescues must be limited to those people who can acquire and maintain the proper skills to use their equipment correctly, who know what knots to use and how to tie them, and who know when they are at their limits of competency. For confined-space rescue, changing from straightforward equipment and tools that require minimal training to rope is an important leap that requires careful study. If the rescuers cannot train on a regular basis and continuously maintain their skills, it would be better to have a plan to utilize outside assistance from a trained team. Even teams that can train on a regular basis may not be able to handle every rescue situation. Knowing who and when to call for help can be the most valuable tool that any rescue team owns. The best advise that this book can offer on the selection and maintenance of ropes and rope equipment is to refer to **NFPA Standard 1983— Fire Service Life Safety Rope and System Components.** This standard is a comprehensive document which is derived from a technical committee of users and manufacturers. It not only covers rope, but also the equipment such as harnesses, carabiners, snap-links, and descending and ascending devices.

The reuse of life safety rope after a rescue is still controversial. NFPA Standard 1983 (1995 edition) provides guidelines regarding the reuse of life safety rope. The manufacturer must provide guidelines on the reuse of rope, and the requirements must include that all of the following conditions be met:

1. Rope has not been visually damaged.
2. Rope has not been exposed to heat, direct flame impingement, or abrasion.
3. Rope has not been subjected to any impact loads.
4. Rope has not been exposed to liquids, solids, gases, mists, or vapors of any chemical or other material that can deteriorate rope.

| NOTE: Tying knots correctly is an invaluable skill, but one that can be lost if you do not practice.

▶ NFPA Standard 1983—Fire Service Life Safety Rope and System Components

an NFPA standard developed for use in the design, testing, use, and maintenance of rope and its associated components used for rescue.

| NOTE: The reuse of life safety rope after a rescue is still controversial.

5. Rope passes inspection when inspected by a qualified person following the manufacturer's procedures both before and after each use.

ANSI defines a **qualified person** as "one who by possession of a recognized degree, certificate, or professional standing, or by extensive knowledge, training and experience has successfully demonstrated the ability to solve or resolve problems relating to the subject matter, the work, or the project." By virtue of rank, a person is not necessarily qualified to inspect a rope or rope equipment.

This set of conditions is very difficult to meet, but consider the consequences if the rope fails. A confined space can contain many different hazards—both physical and chemical. Sharp edges around the opening of the space can damage the rope as can heat within the space or chemicals that might be present. Rope for rescue brings its own set of problems: You must maintain the rope and the rescuers' capabilities of using it. We do not attempt to downplay the value of rope for confined-space rescue, but rather make you realize that as the complexity of your rescue operations increases so do the demands for selecting and maintaining equipment and personnel that will perform as intended.

The equipment used with any life safety rope employed during a rescue should also meet the requirements of the NFPA 1983, including specifications for the original purchase, maintenance, and inspection of the equipment. Carabiners, snap-links, O-rings, and all other related equipment must be maintained so that it is free of corrosion, that parts are operating smoothly (as shown in Figure 11-35), that parts are not deformed for any reason, and that the equipment is free of grease, oil, and other contaminants. You should also have a system for

▶ **qualified person**

as defined by ANSI, a person who by reason of training, education, and experience is knowledgeable in the operation to be performed and is competent to judge the hazards involved and specify controls and/or safety measures.

FIGURE 11-35

Maintaining your equipment in a safe manner is essential to the reliability of the equipment. The duct tape shown here is not an acceptable repair to the damaged lock on this snap hook.

recording equipment inspections and take uncertain or unusable equipment out of service for either testing by the manufacturer or proper disposal. All equipment (harnesses, ropes, carabiners, etc.) must be maintained and inspected on a regular basis, before each use, and after each use. If this equipment has been dropped or impact loaded, it may be unserviceable or require inspection and/or testing before being returned to service.

Consult with the manufacturer for specific recommendations for inspection and testing of equipment. Any equipment that will be discarded should be done so in a fashion that will not allow others to use it. In certain cases, people have found items such as harnesses in the trash and taken them to use in other applications. When the harness failed and people were injured, the previous owner and the manufacturer faced lawsuits. When you dispose of rescue equipment, first make it unserviceable. NFPA 1983 requires that the manufacturer provide information regarding the removal of equipment from service and its destruction.

PEOPLE AND EQUIPMENT

Earlier chapters discussed monitoring equipment, lockout/tagout equipment, and ventilation equipment. All of this equipment is only as good as the people who use it, and those people are limited by what the equipment has been designed to do and how it is maintained. For monitoring devices, personnel must know how to interpret the readings and that the equipment has been calibrated, properly maintained (including batteries), and is reliable. Lockout/tagout equipment is simple enough, but is it periodically inventoried to be sure that you have all of your equipment and that it is functioning properly? As your experience and knowledge base grow, will the equipment you have match your requirements? Ventilation equipment also needs maintenance. The power source for the blowers requires maintenance, including the electrical connections. Damaged or frayed power cords have no place at an emergency scene as they can pose a hazard to rescuers or victims. Cleaning response equipment does more than simply make it shine—it familiarizes personnel with the equipment and provides an in-depth inspection program during the process.

Regardless of your personnel's level of competency, you must work to maintain it. Many times the initial training will provide comprehensive information that is basic to the field and some limited training. Only through repeated practice can emergency personnel hope to develop a working knowledge of the tools and equipment that would be used during a confined-space rescue. That working knowledge can support efficient operations during which the incident commander does not have to give exacting details for the action plan. SOPs are not only supported through training, but are also revised and made applicable by testing them during training and emergency response. Preparing personnel for the tasks they are to be assigned and the risks they will manage is as much of a maintenance plan as it is for any tool or piece of equipment. Part of your maintenance plan for people should be a record of their training as a means of tracking their capabilities.

■ SUMMARY

The resources you have available at a confined-space rescue incident all have limitations. At times the limitations will be built in and inflexible. You will have no choice but to live within the limits. At other times, those limits can be overcome and changed. Early in this book, limiting factors were called *strategic factors,* and that definition has not changed. Before you purchase equipment specifically for confined-space rescue, know what you want to accomplish and then look at the available equipment and buy what will fit your needs. Know what specific limits the equipment has built into it and stay within those limits, thus risk management for your rescuers. You can limit the effectiveness of your equipment by failing to inspect or maintain it, which means that you do not understand the true value of the equipment. Imagine a tripod failing at an emergency because someone lost a locking pin at a drill. People in emergency services know the value of reliability and dependability of their equipment. How many firefighters would enter a hazardous atmosphere wearing SCBA if it could not be relied upon to work correctly? Inspection and maintenance of equipment removes limiting factors by ensuring the availability and usability of the equipment. Having emergency personnel assist in the inspection and maintenance gives them a better knowledge of the equipment.

Although people are not considered equipment, they are certainly resources. Poorly or inadequately trained personnel pose a hazard to the victim, to themselves, and to each other. Before you set out to purchase confined-space equipment, know to what level of capability your personnel must be trained. Define the roles and tasks of personnel to match the appropriate training and equipment. When you have finished your basic training, you have a knowledge set that can be applied over and over again and a level of experience that allows you different ways to apply that knowledge. The experience you gain can come from actual emergencies, but more likely will come from training sessions that reinforce your skills and give you the opportunity to apply them repeatedly. Training is the inspection and maintenance for the competency of your emergency personnel. This inspection and maintenance program is no less important than it is for equipment or other resources.

Have a plan before you begin establishing a confined-space rescue team. Your plan should define the team, its roles, equipment, and maintenance. Work your plan and adjust it as needed. If you cannot afford to maintain your resources, do not waste your money buying them in the first place. You would not drive to the next town without an idea of where you were going or if you thought your car would not make the trip. Why would you do any less when people's lives are at risk and you are asking others to take on some of that risk to help save a life. To quote the well-known leadership specialist, Dr. Stephen R. Covey, "Begin with the end in mind."

NOTE: Training is the inspection and maintenance for the competency of your emergency personnel.

■ REVIEW QUESTIONS

1. An axial load is a load that is transmitted through the axis of the object supporting the load. If a tripod is not loaded axially, how could it fail?

2. Impact loads are a result of the acceleration force being applied by a load in motion. Give an example of an impact load and how the speed of the application of the load can give different results.

3. What is the value of standards when specifying equipment for confined-space rescue?

4. If you have an older piece of equipment that was designed to a recognized standard and an updated version of the standard is created, will your old equipment automatically meet the new standard? Explain your answer.

5. What are the three different classes of harnesses defined in the text? How do they vary in their use and applicability for confined-space rescue?

6. What is the value of having legs that lock to the head of a tripod? Explain your answer.

7. How can you use three lines attached to hoisting equipment to send two rescuers into a confined space? Why would you use three lines instead of two?

8. If you were called to the scene of a confined-space accident and the strategic factors of the emergency proved to be beyond your ability to operate, how could you handle the emergency? Explain how pre-incident information would help you prepare.

9. Why must equipment be inspected periodically, after each use, and before each use?

Glossary

Action limit   the highest percentage of combustible gas that is detected by a combustible gas detector which is considered as the point at which people should leave the area. Generally this is defined as 10 percent of the lower flammable limit. For rescue purposes it may rise to 25 percent of the LFL if the identity of the flammable gas is known, the characteristics of the calibration gas for the combustible gas detector are known, and a relative response can be established between the gas being detected and the combustible gas detector.

Action plan   a plan developed by the incident commander that establishes goals and objectives for the emergency, identifies the resources to be used, and provides the means to accomplish the goals and objectives.

All risk system   an Incident Command or Management System that can be used at many different types of emergencies, including but not limited to fires, police actions, emergency medical calls, and other emergencies that threaten public safety.

American National Standards Institute (ANSI)   an organization that administers and coordinates a voluntary private sector standardization system. Standards developed under ANSI are consensus standards created by representatives from various interest groups.

ANSI Standard A10.14-1991 (Requirements for Life Safety Belts, Harnesses, Lanyards, and Lifelines for Construction and Demolition Use)   a specific ANSI standard that specifies the design, construction, maintenance, and testing of life safety belts and harnesses, among other items.

Atmospheric hazards   conditions which present an atmosphere that can be toxic, flammable, oxygen deficient, oxygen enriched, or that obscures visibility.

Attendant   the person trained and assigned to remain outside of the confined space, monitor conditions inside and outside of the space, and communicate with persons inside. The attendant may not enter the confined space while assigned attendant duties and is expected to call for help if the entrants require rescue. During a confined-space emergency, at least one member of the rescue team should be assigned the duties of attendant and is known as the rescue attendant.

Authorized entrants   those persons trained, assigned, and equipped to enter and work within the confined space. During a confined-space rescue, those persons trained, equipped, and assigned to enter the confined space are known as the rescue entrants.

Axial loads   a load is transmitted through the axis of its supporting device.

Axis   the imaginary line passing through the center of a solid or plane.

Blank flanges   a flange that has no opening in it and is meant to block the flow of a product past the flange.

Blinding   the insertion between two flanges of a device called a blind which has no opening in it and is meant to prevent the flow of a product past the blind.

Blocks   an energy-isolating device meant to stop or obstruct the flow of hazardous energy or products.

Bolted slip blinds   a blinding device that is meant to be bolted directly to a flange to stop or obstruct the flow of a product.

Braided ropes   a type of rope constructed by interweaving the strands of the rope together.

Calibrated   the condition of a measuring instrument after its graduations have been checked or corrected.

Chain   a flexible series of joined links or rings, typically of some type of metal.

Chock   a block or wedge designed to prevent motion of an object it is placed into or under.

Class I, Division I, Group D   electrical equipment specified under the National Electrical Code as meeting particular requirements for safe performance under certain conditions. The designation of class, division, and group refers to distinct hazardous atmospheres that may be present during the use and operation of this equipment.

Class I harnesses   harnesses designed to support a one-person load for escape purposes with the harness fastening around the waist and either under the buttocks or around the legs.

Class II harnesses   harnesses designed to support a two-person load for rescue purposes with the harness fastening around the waist and either around the thighs or under the buttocks.

Class III harnesses   harnesses designed to support a two-person load for rescue purposes with the harness fastening around the waist, either around the thighs or under the buttocks, and over the shoulders to protect against inversion.

Combustible gas indicator (CGI)   a metering device intended to detect and measure the presence of a flammable gas based on how close the gas is to the lower flammable limit of the calibration gas.

Command post   the location from which all incident activities are directed by the incident commander.

Communications   the act of sending and receiving a message and having the message understood by the receiver. Effective communications may require the use of equipment to facilitate carrying the message.

Confined-space supervisor   the person assigned responsibility for ensuring that the requirements of a confined-space program have been met prior to, during, and after entry by persons into a confined space. During a confined-space rescue operation, either the incident commander or specific person such as the safety officer should be assigned this duty.

Construction   the materials from which the confined space is built. As a strategic concern, the materials of construction may contribute to the limiting factors presented by the confined-space emergency.

Contents   for confined-space rescues, the materials within the confined space. These contents may be either gas, liquid, or solid and may or may not contribute to the limiting factors affecting the confined-space emergency.

Corrosive   a material that can be acidic or basic and, because of those properties, can damage human skin or rapidly corrode metal.

Decibel   a unit for measuring sound intensity.

Defensive actions   actions taken in which there is no intentional contact with the hazards of a confined space. Rescue from outside of a confined space by retrieving the victim with in-place retrieval equipment is one example.

Degradation   the reduction of protective properties of chemical protective clothing by mechanical, thermal, or chemical means with a loss of integrity of the garment.

Destructive tests   a means of testing during which the test item is tested to failure.

Direct reading instruments   detection and monitoring instruments that give a reading based on a graduated scale.

Disconnect switches   an electrical switch designed to isolate the electrical source from the equipment that it powers by disconnecting the power supply from the equipment.

Dynamic rope   a type of kernmantle rope that is intended to stretch under load. This stretching may be designed to provide absorption of a shock load.

Eccentric loads   a load applied so that the force of the load is off center of the supports carrying the load.

Engulfment   the surrounding and effective capture of a person by a liquid or finely divided (flowable) solid substance that can be aspirated to cause death by filling or plugging the respiratory system or that can exert enough force on the body to cause death by strangulation, constriction, or crushing.

Entrant   see *Authorized entrants.*

Entry permit   a written document which must be completed prior to entry into a confined space and defines the hazards of the space, the precautions to be taken, the type of work that will be performed, the roles of personnel involved in the entry, and other specific details.

Exposures   the people, property, and systems that may be affected by the confined-space rescue operations.

Flammable or **explosive range**   a definite concentration of flammable vapors in air, for a particular material, at which combustion will occur. There is a lower flammable limit and an upper flammable limit at which the vapors are either too lean to burn or too rich.

Flash point   the minimum temperature at which a material will produce enough vapors, in air, to form an ignitable mixture near the surface of the material.

Hazard and risk assessment   determining what has happened to create the emergency, what conditions are still present or will evolve at the emergency, and then predicting what can be done to resolve the emergency.

Hydrogen sulfide   a flammable, toxic gas that can be created by the decomposition of organic materials.

Hypothermia   lowering of the body's core temperature.

Immediately dangerous to life and health (IDLH)   the maximum level to which one could be exposed and still escape without experiencing any effects that may impair escape or cause irreversible health effects.

Impact loads   a load applied in a very short duration so as to include the effects of acceleration in the load.

Incident command system (ICS)   a recognized system for providing management of personnel, resources, and activities during emergency operations.

Incident management system   a recognized system for providing management of personnel, resources, and activities during emergency operations.

Incident priorities   the order of precedence given to the most basic goals of life safety, incident stabilization, and property conservation during an emergency operation.

Inerting   the introduction of an inert gas or a gas which will not support combustion into a tank or vessel so as to exclude oxygen from the tank or vessel.

Kernmantle ropes a type of rope made with a load-bearing core (kern) and an outer sheath or braided cover (mantle). There are both static and dynamic types of kernmantle rope.

Laid rope a type of rope that is made by twisting smaller strands of rope together.

Latch a fastening device that consists of a bar that falls into a notch to prevent opening or operation of the object it secures.

Lead time the amount of time that is required for a specific action to occur after notification is made of the need for that particular action. An example of lead time would be the amount of time that would be necessary to set up a retrieval system after it was determined that a retrieval system was needed.

Level A the highest level of skin and respiratory protective equipment for exposure to hazardous materials. Typically this level of protective clothing is designed to be vaportight to protect against gases or vapors.

Level B the second highest level of skin and respiratory protective equipment for exposure to hazardous materials. Typically this level of protective clothing is designed to be liquidtight to protect against splash hazards.

Level C the third level of skin and respiratory protective equipment for exposure to hazardous materials. Typically this level of protective equipment is designed to be used when the hazards are known, monitored, and controlled. Because this level of protective equipment uses an air-purifying respirator, it is not intended to be used for emergency response.

Level D the lowest level of skin and respiratory protective equipment for exposure to hazardous materials. This level provides no protection and typically consists of street clothing and no respiratory protective equipment.

Liaison within the incident command system, the command staff position responsible for establishing and maintaining interaction with other agencies required to handle the incident.

Life hazard as a strategic factor during emergency response, the threat posed to the victims, emergency responders, and spectators.

Line valves valves in the piping which allow product to travel into a tank, vessel, vat, or other confined space. Line valves may be useful in locking out the flow of a product or other energy source into the space.

Location and accessibility as a strategic factor, the physical location of the confined space and the available means of gaining entry to it.

Lockout/tagout elimination and control of hazardous sources of energy or products.

Logarithmic scales a scale, such as the pH scale, in which a change between whole numbers represents an exponent of the power to which the change will be raised. For example, a material with a pH of 5 is 10 times as acidic as a material with a pH of 6, whereas a material with a pH of 4 would be 100 times as acidic as a material with a pH of 6.

Manually operated electrical circuit breakers circuit breakers within an electrical circuit that are designed and intended to be safely operated by direct manual manipulation.

Medical monitoring basic medical evaluation of emergency response personnel by determining and recording basic vital signs such as blood pressure, respirations per minute, and pulse. Medical monitoring is typically conducted prior to and after entry to an incident to determine if any significant changes have taken place.

Methane a colorless, odorless, flammable gas consisting of one carbon atom and four hydrogen atoms, also called natural gas.

National Fire Protection Association (NFPA) an international nonprofit organization advocating scientifically based consensus codes and standards, research, and education for fire and related safety issues.

National Institute for Occupational Safety and Health (NIOSH) the agency within the U.S. Department of Health and Human Services that identifies work-related diseases and injuries and the potential hazards of new work-related technologies and practices.

Negative-pressure ventilation the systematic removal of air, gases, and other airborne contaminants by using a fan to draw the air or contaminants out of the confined space.

NFPA Standard 1983 - Fire Service Life Safety Rope and System Components an NFPA standard developed for use in the design, testing, use, and maintenance of rope and its associated components used for rescue.

Nondestructive tests a method of testing in which an item is tested within specific parameters to determine if it can meet the requirements of the test and recover within acceptable limits. The test item can be returned to use if it meets the requirements of the test. An example is hydrostatic testing of compressed gas cylinders.

Non-permit confined space confined spaces that do not contain or, with respect to atmospheric hazards, have the potential to contain any hazard capable of causing death or serious physical harm.

Occupational Safety and Health Administration (OSHA) the federal agency within the U.S. Department of Labor that is responsible for creating and enforcing workplace safety and health regulations.

Offensive actions actions taken in which there is expected to be intentional contact with the hazards of a confined space. Rescue by entering the confined space to retrieve the victim is one example.

OSHA Standard 1910.146 Permit-required confined spaces the specific section of the code of Federal Regulations which regulates confined spaces and the manner in which activities can occur within those spaces.

Parts per million (PPM) the number of units of a particular material occurring in a total volume of one million units. One part per million is the equivalent of 1/10,000 of 1 percent.

Penetration the reduction of protective properties of chemical protective clothing which can occur due to zippers, seams, and other openings in the protective clothing.

Permeation the reduction of protective properties of chemical protective clothing caused by the movement of the contaminant on a molecular level.

Permissible exposure limit (PEL) a time weighted average (TWA) concentration that must not be exceeded during any 8-hour work shift of a 40-hour workweek. A short-term exposure limit (STEL) is measured over a 15-minute period.

Permit-required confined space confined spaces that meet the definition of a confined space and have one or more of these characteristics: (1) contain or have the potential to contain a hazardous atmosphere, (2) contain a material that has the potential for engulfing an entrant, (3) have an internal configuration that might cause an entrant to be trapped or asphyxiated by inwardly converging walls or by a floor that slopes downward and tapers to a smaller cross section, and (4) contain any other recognized serious safety or health hazards.

Personal Alert Safety Systems (PASS) a device, typically worn by firefighters, intended to provide a means of automatically sounding an alarm when the wearer remains motionless for more than 30 seconds or manually by the wearer if in need of assistance.

pH paper a basic detection device consisting of pH-sensitive paper that will change colors when exposed to an acid or base. Some pH papers are designed to change color in proportion to the pH level of the material to which they are exposed.

pH pen a monitoring device, resembling a pen, which gives a direct reading of the pH of the material to which it is exposed.

Physical hazards hazards within a confined space which are produced by mechanical, electrical, chemical, or thermal means and endanger personnel in the confined space.

Positive-pressure supplied air respirator (SAR) a form of respiratory protection in which the self-contained air supply is remote from the wearer, the air is supplied to the wearer by means of an air hose, and the pressure within the facepiece is greater than the surrounding atmospheric pressure.

Positive-pressure ventilation the systematic removal of air, gases, and other airborne contaminants by using a fan to blow into a space to push the air or contaminants out.

Public Information Officer within the incident command system, the command staff position responsible for providing the press and media with information about the incident as authorized by the incident commander.

Quad Pod™ a brand name for a retrieval support device consisting of a davit arm centered over four legs which support the device.

Qualified person as defined by ANSI, a person who by reason of training, education, and experience is knowledgeable in the operation to be performed and is competent to judge the hazards involved and specify controls and/or safety measures.

Radio frequency (RF) interference electromagnetic interference caused by a signal generated by an electrical device.

Resource management the allocation and maintenance of resources required during an emergency.

Resources the people, supplies, and equipment required during an emergency.

Saddle Vent™ a brand name for a ventilation device meant to be placed in the manway opening that will allow air to be directed through the opening with minimal obstruction to the manway for the movement of people.

Safety Officer within the incident command system, the command staff position responsible for incident safety including identifying potentially hazardous situations and the enforcement of safety procedures and safe practices.

Self-contained breathing apparatus (SCBA) a form of respiratory protection in which the self-contained air supply and related equipment is attached to the wearer and the pressure within the facepiece is greater than the surrounding atmospheric pressure.

Single command a form of command within the incident command system when a single individual is responsible for the tasks assigned to the incident commander.

Special problems as a strategic factor, this is a broad category for problems which are unique to a particular incident and would only be likely to recur on an infrequent basis.

Staging of resources   a method of managing resources in which those not actively being used are kept or staged in a particular area in preparation for use.

Standard 1670- Standard on Operations and Training for Technical Rescue   a proposed NFPA standard currently being developed to define and categorize technical rescue qualifications.

Standard operating procedures (SOPs)   written guidelines for handling a specific situation.

Static load   a load applied when the load is at rest.

Static rope   a type of kernmantle rope that has little stretch when under load.

Time   as a strategic factor, the time of day, the day, week, or year and the relative impact that it will have on emergency operations.

Transformer retrieval support   a device which is designed to be bolted to the flange around a manway opening for vertical entry.

Tripod   three-legged retrieval support device.

Unified command   a form of command within the incident command system when more than one individual is responsible for the tasks assigned to the incident commander. Unified command allows for a single set of goals and objectives and a single plan of action to be developed.

U.S. Environmental Protection Agency (EPA)   the federal agency responsible for developing and enforcing environmental regulations.

Vapor density   the weight of a given volume of gas or vapor compared to an equal volume of air at the same temperature and pressure. Air has a vapor density of 1. Gases with vapor densities less than 1 are lighter than air, whereas vapor densities greater than 1 are heavier than air.

Ventilation   the systematic removal and replacement of air and gases within a space.

Victim recovery   the recovery of a victim who is known or expected to be deceased.

Weather   as a strategic factor, the effects that can be expected due to temperature, wind, precipitation, and other climatic factors.

Working loads   the maximum weight that a rope is expected to support. For a one-person rope that weight is 300 pounds and for a two-person rope that load is 600 pounds.

Wristlets   straps designed to be placed around the wrists of a person to allow him to be raised or lowered through a vertical opening while hanging from the straps.

Index